The International Library of Psychology

THE HANDS OF CHILDREN

Founded by C. K. Ogden

The International Library of Psychology

INDIVIDUAL DIFFERENCES
In 21 Volumes

I	The Practice and Theory of Individual Psychology	*Adler*
II	The Neurotic Constitution	*Adler*
III	Duality	*Bradley*
IV	Problems of Personality	*Campbell et al*
V	An Introduction to Individual Psychology	*Dreikurs*
VI	The Psychology of Alfred Adler and the Development of the Child	*Ganz*
VII	Personality	*Gordon*
VIII	The Art of Interrogation	*Hamilton*
IX	Appraising Personality	*Harrower*
X	Physique and Character	*Kretschmer*
XI	The Psychology of Men of Genius	*Kretschmer*
XII	Handreading	*Laffan*
XIII	On Shame and the Search for Identity	*Lynd*
XIV	A B C of Adler's Psychology	*Mairet*
XV	Alfred Adler: Problems of Neurosis	*Mairet*
XVI	Principles of Experimental Psychology	*Piéron*
XVII	The Psychology of Character	*Roback*
XVIII	The Hands of Children	*Spier*
XIX	The Nature of Intelligence	*Thurstone*
XX	Alfred Adler: The Pattern of Life	*Wolfe*
XXI	The Psychology of Intelligence and Will	*Wyatt*

THE HANDS OF CHILDREN

An Introduction To Psycho-Chirology

JULIUS SPIER

With an Appendix on the Hands of the Mentally Diseased by Herta Levi

ROUTLEDGE

First published in 1944 by
Routledge

Reprinted in 1999 by
Routledge
2 Park Square, Milton Park, Abingdon, Oxon, OX14 4RN

Transferred to Digital Printing 2007

Translated by Victor Grove, Drawings by F W W Simon

British Library Cataloguing in Publication Data
A CIP catalogue record for this book
is available from the British Library

The Hands of Children
ISBN 0415-21068-2
Individual Differences: 21 Volumes
ISBN 0415-21130-1
The International Library of Psychology: 204 Volumes
ISBN 0415-19132-7

PREFACE

It was the intention of Mr. Julius Spier to follow up this volume with two more works, " The Hands of Adults " and " The Hands of the Mentally Diseased." The author, however, died suddenly a few months ago and hence was deprived even of the satisfaction of witnessing the publication of his first book, which was to be his last. Thus a number of alterations and additions contemplated by him cannot be introduced, and owing to war-time difficulties I did not receive the original prints of the thirty hands of children which he had intended to include in this book. As some of the photographic copies at my disposal are unsatisfactory I had no choice but to omit seven prints and analyses.

In the name of the author I avail myself of this opportunity to express my sincere gratitude to Dr. V. Grove, who is responsible for the careful and sympathetic English version of this book and whose assistance has enabled me to secure its publication.

Again and again those who knew of Spier's great reputation on the Continent, which was based on innumerable lectures, courses, as well as on an extensive practice, naturally asked, why in all those years he had never published anything except occasional articles and a small pamphlet for the use of his pupils. The reason for Spier's reluctance in spite of an ever-growing demand, and his own realisation of the pressing need for a comprehensive work on Psycho-Chirology, was his resolve conscientiously and critically to sift his material, and empirically to test his theories exhaustively, before submitting them to the public.

One of his last utterances before he died, " but my work is not yet finished ", should be continually borne in mind by those who knew his work, and above all by us who are his pupils. They impose upon us the obligation to keep his work alive and to carry on his achievements by further critical study and scientific research. For at present Chirology is being re-born ; it stands at the gate of modern development, and Julius Spier was one of its pioneers and greatest of its workers.

It is certain that the spirit of his work will live on among his pupils and the thousands of people who passed through his consulting-rooms ; and with it the memory of a man who had vision, who worked with the conscientiousness of conviction, and, above all, was a creative mind. Although he could not live to "finish his work", the task he set himself has been more than accomplished, and it is to be hoped that this book, which later on will be supplemented from his unpublished writings, will considerably widen and stimulate the circle of those sincerely interested in a branch of scientific knowledge which is still exposed to so much controversy.

HERTA R. LEVI.

LONDON,
February 1943.

PREFACE
TO THE SECOND EDITION

When editing the first edition of *The Hands of Children* I was aware of the fact that Julius Spier had intended it to be the forerunner of a series of chirological books, one of which was to be The Hands of the Mentally Diseased. As he died before the publication of his first book, I had hoped to be able to supplement his work later on from the numerous notes and prints which he had accumulated in the course of his extensive practice. But I have been unable to carry out my intention, for the last war, and the complications following in its wake, rendered it impossible for me to secure his notes. In the meantime, however, I had been given the opportunity of compiling a fair amount of interesting chirological material at a mental hospital in this country, the staff of which afforded me generous facilities. I hesitated to publish this collection of prints of mental cases, for I felt that it would be premature to draw definite conclusions or to lay down hard and fast rules. On the other hand, it seemed a pity to waste valuable prints which could help to further chirological research, for what is needed most urgently at the present juncture is more, and again more, material. As a scientific method Chirology is still in its infancy and requires a good deal of body-building nourishment in the shape of factual data and statistics. Without it, chirologists are sometimes inclined to jump to conclusions, or to simplify matters unduly, or to overstate their case when they should rather understate it until sufficient statistical evidence, compiled with the greatest accuracy, is made available. Very few of us have the great intuitive powers of a Julius Spier, and therefore have to rely more on scientific methods and exchange of information.

I hope that this material will offer many interesting points of comparison to the student of chirology and will help to round off the book. Particularly the last plate links up with the first, showing the hands of a man of seventy who in his second childhood became a victim of senile dementia, and those of a mentally deficient boy. I trust that by including this Appendix on The Hands of the Mentally Diseased I have at least partly fulfilled the promise made in my preface to the first edition.

H. R. L.

January 1955

CONTENTS

Preface to the First Edition — *page* v
Preface to the Second Edition — vi
Introduction *by* Dr. C. G. Jung — xv

THE THEORETICAL FOUNDATION OF PSYCHO-CHIROLOGY

1. Why the Hands of Children? — 1
2. General Remarks about Psycho-Chirology — 4
3. The Outer Hand — 10
4. The Fingers — 18
5. The Position of the Hand — 35
   The Relationship of the Fingers towards Each Other — 35
   The Peculiarities of Each Finger — 35
6. The Significance of the Right and Left Hands — 47
7. The Mounts — 58
8. The Main Lines in the Palm of the Hand — 66
9. Subsidiary Lines and Various Other Features of the Inner Hand — 91
10. Practical Notes — 97
11. Chirology Applied to the Hands of Children — 106

Figures — 110
Appendix One: Specimen Analyses of Hands of Children — 159
Appendix Two: The Hands of the Mentally Diseased — 167
   Notes on Hands of Schizophrenics — 172
   Notes on Hands of Non-Schizophrenics — 179

Table One 184
Table Two 185
Postscript 187
Index 189
Index of Figures 199
Plates *at end*

PLATES

THE CHILDREN'S HANDS

I. The Hands of a Boy of Twelve

II. The Hands of a Boy of Nine

III. The Hands of a Boy of Sixteen

IV. The Hands of a Boy of Fourteen

V. The Hand of a Boy of Fourteen

VI. The Hand of a Boy of Twelve

VII. The Hands of a Boy of Six

VIII. The Hands of a Woman of Forty

IX. The Hands of a Girl of Three

X. The Hand of a Boy of Four

XI. The Hand of a Boy of Fifteen

XII. The Hands of a Girl of Thirteen

XIII. The Hands of a Boy of Thirteen

XIV–XVI. The Hands of a Boy and his two Sisters

XVII–XVIII. The Hands of a Girl at the Age of Twelve and Nineteen

XIX–XX. The Hands of a Boy at the Age of Twelve and Eighteen

XXI–XXII. The Hands of Two Fraternal Twins at the Age of Twelve

XXIII. The Left Hand of the Brother of the Twins, aged Seventeen

THE HANDS OF MENTALLY DISEASED PEOPLE

XXIV. The Hands of a Case of Primary Dementia. Male. Age Forty-six

XXV. The Hands of a Case of Simple Schizophrenia. Male. Age Eighteen

XXVI. The Hands of a Case of Primary Dementia. Male. Age Twenty-nine

XXVII. The Hands of a Case of Primary Dementia. Female. Age Thirty-eight
XXVIII. The Hands of a Catatonic Schizophrenic. Male. Age Thirty-three
XXIX. The Hands of a Catatonic Schizophrenic. Male. Age Thirty-seven
XXX. The Hands of a Catatonic Schizophrenic. Male. Age Twenty-two
XXXI. The Hands of a Schizophrenic. Female. Age Fifty
XXXII. The Hands of a Schizophrenic. Female. Age Sixty-five
XXXIII. The Hands of a Schizophrenic. Female
XXXIV. The Hands of a Schizophrenic. Female. Age Thirty-seven
XXXV. The Hands of a Schizophrenic. Female. Age Forty-four
XXXVI. The Hands of a Schizophrenic. Female. Age Twenty-three
XXXVII. The Hands of a Schizophrenic. Female. Age Forty-three
XXXVIII. The Hands of a Schizophrenic. Female. Age Forty-three
XXXIX. The Hands of a Schizophrenic. Female. Age Forty
XL. The Hands of a Case of Early Schizophrenia. Male. Age Fifteen
XLI. The Hands of a Schizophrenic. Male. Age Thirty-four
XLII. The Hands of a Schizophrenic. Female. Age Fifty-five
XLIII. The Hands of a Schizophrenic. Female. Age Thirty-four.
XLIV. The Hands of a Schizophrenic. Female. Age Thirty-one
XLV. The Hands of a Schizophrenic. Female
XLVI. The Hands of a Schizophrenic. Female. Age Forty-five

XLVII. The Hands of a Schizophrenic. Female. Age Fifty-four

XLVIII. The Hands of a Schizophrenic. Female. Age Forty-four

XLIX. The Hands of a Schizophrenic. Female. Age Fifty-two

L. The Hands of a Schizophrenic. Female. Age Thirty-seven

LI. The Hands of a Schizophrenic. Female. Age Twenty-six

LII. The Hands of a Paranoiac. Male. Age Fifty-one

LIII. The Hands of a Paraphreniac. Female. Age Thirty-five

LIV. The Hands of a Case of Manic-Depressive Psychosis. Male. Age Sixty-two

LV. The Hands of a Case of Manic-Depressive Psychosis. Male. Age Sixty-eight

LVI. The Hands of a Case of Mania. Female. Age Fifty-seven

LVII. The Hands of a Case of Mania. Male. Age Sixty-seven

LVIII. The Hands of a Case of Psychoneurotic Depression. Female. Age Twenty-three

LIX. The Hands of a Case of Uncertain Diagnosis. Female. Age Forty-five

LX. The Hands of a Case of Uncertain Diagnosis. Female

LXI. The Hands of a Case of Uncertain Diagnosis. Female. Age Twenty-two

LXII. The Hands of a Doubtful Case. Male. Age Twenty-five

LXIII. The Hands of a Case of Bromide Intoxication. Male. Age Thirty-one

LXIV. The Hands of a Case of Reactive Depression. Female. Age Forty-nine

LXV. The Hands of a Case of Involutionary Melancholia. Female. Age Sixty-four

LXVI. The Hands of a Case of Involutionary Melancholia. Female. Age Forty-nine

LXVII. The Hands of a Case of Involutionary Melancholia. Female. Age Sixty-two

LXVIII. The Hands of a Case of Involutionary Melancholia. Female. Age Fifty-seven

LXIX. The Hands of a Case of Senile Dementia. Male. Age Seventy

INTRODUCTION

by

C. G. JUNG

Chirology is an art which dates back to very ancient times. The ancient physician never hesitated to make use of such auxiliary systems as chiromancy and astrology for diagnostic and prognostic purposes as is shown, for instance, by the book of Dr. Goclenius who lived at the end of the sixteenth century in Würzburg. The rise of the Natural Sciences and with it of rationalism in the eighteenth century were responsible for the contemptible treatment and defamation of these ancient arts which could pride themselves on a thousand years of history, and this led to the rejection of everything which on the one hand defied a reasonable explanation and verification by experiment or, on the other, made too exclusive a claim on intuition. On account of the uncertainty and paucity of scientific knowledge in the Middle Ages, even conscientious thinkers were exposed to the danger of applying their intuition rather to the promotion of superstition than science. Thus all early, and particularly medieval, treatises about this subject are an inextricable tangle of empiric and phantastic facts. To establish a scientific method and to obtain reliable results it was necessary, first of all, to make a clean sweep of all these irrational methods. In the twentieth century, however, after two hundred years of intensive scientific progress, we can risk resurrecting these almost forgotten Arts which have dragged on a despised existence in semi-obscurity; and we can risk, testing them in the light of modern knowledge for possible truths.

The totality-conception of modern biology which is based on the evidence of a host of observations and research does not exclude the possibility that *hands*, whose shape and functioning are so intimately connected with the psyche, might provide revealing and, therefore, interpretable expressions of psychical peculiarity, that is, of the human character. Modern science increasingly relinquishes the medieval conception of the separateness of body and mind, and just as the body in the view of science is neither something mechanical nor chemical, so the mind seems to be but another aspect of the living body. Conclusions relating to one or the other would therefore seem to be within the range of scientific operation.

I have had several opportunities of observing Mr. Julius Spier at work, and must admit that the results he has achieved have made a lasting impression on me. His method, though predominantly intuitive, is based upon a vast practical experience. Experiences of this kind can be rationalised to a great extent, that is to say, they admit of reasonable explanation once they are available. The manner, however, in which they are obtained depends, apart from routine, in its most decisive points, upon a subtly differentiated creative intuition which in itself implies individual specific talent. We, therefore, can hardly expect persons with nothing but average intelligence to be able to master this method. There is, however, a definite possibility that people who are intuitively gifted should be able to achieve similar results provided they are properly taught and trained. Intuition is not just an isolated talent but a regularly occurring function which is capable of development. Like the function of seeing and hearing, it has its specific field of experience and a specific range of cognition.

The findings and knowledge expounded in this book are of essential importance for psychologists, doctors, and educationalists. Spier's Chirology is a valuable contribution to character-research in its widest application.

The Theoretical Foundation of Psycho-Chirology

CHAPTER ONE

WHY THE HANDS OF CHILDREN?

Become what thou art.
PINDAR.

I have chosen Pindar's discerning words because they contain the starting-point as well as the aim of my book, which is not only to interpret and to analyse but also to point out the road of development which leads towards perfection, and to achieve the true synthesis of the personality.

I base my approach on the assumption that man is born with definite dispositions which in normal cases determine his intellectual and spiritual development at least in part, though they may be decisive in abnormal cases. The essential factor is a person's attitude to the events and phenomena of his life, and here the dispositions of a person normally present a number of possibilities. Most people, however, because of wrong education and haunting impressions from childhood which have produced disturbances and inhibitions, are unable to adopt that attitude to life which is in accordance with their actual dispositions. The ignoring and neglecting of their true dispositions have forced them into an unrelated, uncongenial condition and warped their outlook on life. This restricted attitude places them in social isolation and egocentric confinement, causing all kinds of disturbances, inhibitions and conflicts and often brings about a complete standstill in the development of the personality. It is at this point that the work of the chirologist begins. Apart from ascertaining the character, the talents, the unconscious influences, the effect of the milieu, it is his task to find out how and by what a person is prevented from leading a life according to his own natural dispositions, how and why a person is again and again compelled to go through certain experiences which bind and limit him, and impede his true inner development. In this clarifying process it is necessary to show him in what way he is connected with his ancestors, that is to say, to make him realize that his life and his destiny are directly connected not only with his dispositions but also with the life experiences of his forefathers.

The mechanisation and mad speed of the present day have driven man further than ever before into a state of social isolation which makes him forget that he is not merely an individual but also a part of a much greater whole. The conscious realisation that his reactions and experiences are closely linked with the life of his forbears lends a new meaning to his life and often gives it an inner significance, breaking down his artificial isolation and leading him from his exaggerated egocentric attitude to a better relationship with and adaptation to the community. He finds his proper place in the universe and he is able to bear all his sufferings and hitherto incomprehensible difficulties consciously and, therefore, with greater ease. He learns to accept what up to now he had resisted and he is now able to throw overboard the negative ballast which had burdened his mind and prevented him from the positive realisation of his true dispositions, so that at last he can become what he is.

In the case of most adults the task of the psycho-chirologist is necessarily of a therapeutic nature ; with children, however, it is possible to intervene in time and to take prophylactic steps.

From the first year the hand of a child clearly shows its dispositions with all their subsidiary traits, particularly the right hand, which according to my observations must be looked upon as the ancestral hand. If the hand is not clearly lined and marked in early childhood, it will be so at the latest between the eighth and twelfth years, which will then enable the experienced chirologist to determine the dispositions of the child, in which direction they tend, by what means they can be furthered and what may be preventing their natural development. Every normally affectionate mother and every loving father will look upon their child as their most precious possession, and yet how do they treat—or rather ill-treat—it. They usually bring it up according to the negative experiences of their own childhood and consequently go to the opposite extreme. They are inclined to spoil and pamper the child if they discover a positive trait of their own in it, and to oppose and discourage it whenever the child is " disappointing " in this respect. The parents mostly seek compensation for all that had been denied to them and thus treat the child accordingly, and burden it with their own personal problems and conflicts. It is indeed a rare occurrence for parents to guide and educate their child without self-love and according to its own dispositions and talents. It is, however, extremely difficult to discover the true dispositions and talents of a child, particularly if the parents have very little or no knowledge of their own selves. So many parents never become what they should have become, and unconscious of this fact, try to find the fulfilment of their own frustrated wishes in their child without realising that they are continuing the vicious circle which now prevents their own child from finding the fulfilment of its true possibilities and talents. Here the analysis of the hand offers a unique possibility of solving both difficulties : to ascertain the disposition of the parents, and particularly that part of the disposition with which the child is directly, biologically and psychologically, connected, and the individual disposition of the child. A chirological analysis very often shows that the difficulties in the development of the child are not produced by its own inadequate dispositions but by conflicts and mostly unconscious difficulties of the parents which disturb, upset and frighten the child or force it into opposition and obstinacy.

In the concluding practical section of this book (see Appendix, pp. 159–166) I shall analytically demonstrate by means of 23 prints of hands to what extent the psychology of children is expressed in the lines of their hands and how the destinies of children are sometimes conditioned and influenced by parents. These short analyses will show how difficult and hard it is to become what we essentially are, and help to achieve the aim of this book : to make a small contribution towards the solution of this problem which so vitally concerns the inner meaning of our existence.

CHAPTER TWO

GENERAL REMARKS ABOUT PSYCHO-CHIROLOGY

As far as acknowledgment and serious appreciation are concerned, chirology is still the Cinderella among all expressional systems, because it is still exposed to the odium of fortune-telling and charlatanry. The main reason for this may be that the general public as well as doctors, teachers, educationalists and psychologists confuse chirology with chiromancy and it is, therefore, essential that the position should be clarified.

Chirology deals with the shape of the hand, the fingers and nails and the lines of the palm. By an analysis and interpretation of all the features of the hand the chirologist endeavours to construct a complete psychological character-portrait of the personality, taking into account life experiences, dispositions, talents, vocational aptitude and dispositions to illnesses. The first and fundamental commandment of chirology is that no single sign, trait or line shall be considered and valued on its own strength but only in relation to all the features concerned.

The devotee of chiromancy, however, draws his conclusions about the qualities and character-dispositions of a person from definite signs, without considering the dynamic and psychological factors. Thus on the whole he ascertains external events and circumstances of life from certain signs in the hand without connecting these events with certain qualities of the character. Chiromancy, therefore, on the whole, deals with the prediction of the future. For the layman it is often rather difficult to separate chirology from chiromancy, for many hand readers style themselves chirologists but still indulge in the old manner of interpreting isolated traits, that is to say, they predict the future.

Various scholars, doctors and important scientists of the past have occupied themselves with and written about chirology scientifically. I only mention a few such as Aristotle, Savonarola, Paracelsus, Lavater in his Physiognomy, and most comprehensively, the doctor and philosopher Gustav Carus in his Symbolism of the Human Form. Between 1650 and 1730 chirology was officially included in the curriculum of the German universities of Leipzig and Halle.[1]

In my method, which is based on my independent observations and thirty years of practical experience, I have incorporated hardly more than the four main lines of the old chiromantic systems, and it differs fundamentally from all extant chirological systems because it is not concerned with the static features alone but also, and mainly, with the dynamic aspect. I also attribute great importance to the consideration of a person's forefathers in the analysis of the hand, biologically as well as psychologically.

The human soul is not a finished and unalterable entity but is in a state of continuous flux. Human beings have dispositions and talents which according to the influence of their environment, education, and other conscious and unconscious impressions, can be furthered and developed, or inhibited and suppressed, and it is the task of the chirologist to ascertain the original dispositions of a person and to what extent the above-mentioned influences and other alien forces have prevented their free development. Thus my work is not only concerned with the mere description of a person's present state of mind but also with the awakening of latent and buried possibilities and the indication of the true tendency of his inner being.

As I have already pointed out in the previous chapter, to understand and appreciate a personality in its entirety, it is necessary to consider certain aspects of the history of the person's family as well as all the psychological and biological facts connected with it. The chirological approach to this is provided by the right hand, the ancestral hand. Its relationship to the left hand, which is expressive of the individual personality, will be discussed in greater detail in a later chapter.

In my chirological practice I lay such great stress on the past because it is the past which makes us understand the present and because the true understanding of the present and its purposeful adjustment to the whole personality are the condition for a positive and constructive development. As I attribute great importance to all experiences and impressions received in the past, it is only logical that I should strictly refuse to make pronouncements about the future, for if I did describe future events I should automatically exclude the active participation

[1] Those interested in the history of chirology will find a detailed account in R. Engelhard's book on the subject.

of a person in the shaping of his own development. By telling a person correct things about his character and his past, that is to say things he is able to verify, we tend to suggest an atmosphere of confidence which can remain equally effective in the case of subsequent negative predictions and the person may thereby be induced to concentrate with all his strength on these bad future possibilities. This attitude has a paralysing effect upon the natural powers of resistance and at the same time deprives the person of his inner possibilities of escape from what he now believes to be his doom. The fact that he, consciously or unconsciously, concentrates all his attention upon the "unavoidable catastrophe," actually drives him into it, and thus the prediction comes true.

In this connection I should like to elaborate the statement made cursorily in the first chapter that the normal person is equipped with a number of dispositions which open out various possibilities as he deals with the facts and events of life. The impact, however, with the environment and its different individualities, particularly with those of the milieu, often causes these dispositions to be forced into a one-sided direction and to assume a one-sided attitude. Consequently the freedom of scope for action and with it the freedom of the will in its attitude to life are impeded and limited. Thus it is the task of the conscientious and responsible chiro-analyst to restore the original variety of dispositions and with it the possibility of a free choice. The chiromantic hand-reader and predictor of future events works in the opposite direction and, therefore, produces the opposite, negative result which we have noticed above.

It would, however, be wrong to assume that only negative predictions can have a blighting effect on a person's powers of development. Let us assume that a man is told of a great external success which is coming to him in the near future. Sure of success, he might be induced to give up working to perfect his own personality or mode of life, and to let himself go, thus doing harm to his inner self. Besides, such predictions cannot be made with an adequate measure of certainty because the lines and signs of the inner hand upon which the predictions are based are subject to changes, they may disappear within longer or shorter intervals or new ones may emerge. Moreover the signs of the inner hand have not only an external but also an inner reference. The fortune-teller who is only concerned with static phenomena and is bound to correlate his findings mainly with external events, will shackle and restrict his consultee to a fixed attitude. Inner processes, however, to an even greater extent than external events, are especially dependent on a free display and interplay of energies and forces.

An important factor in this connection is the general trend of our age, and of the individual, towards materialism, which means that the responsible psycho-chirologist whose duty it is to give his client greater inner stability should not foster and emphasise that tendency. He should realise that the restlessness and inner difficulties and unhappiness of so many people are to be attributed to the contagious materialist spirit of our period. The resulting inner isolation induces them to seek the advice of the chiro-analyst ; without being aware of it, a conflict has arisen between their vague but genuine striving for cognition, for inner development, and the contagious tendency of the age with which they are not in harmony. A mere prediction of future events would only increase their dilemma and inner isolation because it would fail to show the interconnectedness of inner and external life. Though the chirologist cannot and should not make predictions, he can, however, discover and point out possibilities of development and give prognostic advice in case of difficult, pathological or psycho-pathic dispositions. In such cases the chirologist bases his conclusions about possible future developments upon certain unalterable signs and verifiable symptoms.

The decisive factor in the treatment of and the correct attitude to these questions is, after all, the personality of the chirologist. As he is dealing with problems of education, psychotherapeutic treatment and analyses of children and adults, and thus has to shoulder considerable personal responsibility, it should be made a fundamental condition and minimum demand that he should know his own character very well and, above all, be conscious of all his weaknesses and imperfections. A chiro-analyst with a sense of moral responsibility should know himself to be free of lust for power, sadism, personal embitterment and resentment, any of which might lead him to repudiate and deny the world of values. I wonder how many disastrous predictions have been made because of an unconscious sadist trait of the predictor, and how many destructively negative character analyses must be attributed to a negative attitude of the analyst. It would, of course, be equally wrong to exclude and suppress negative traits in a graphological or chirological analysis, but they should be presented in a constructive way, for the activity of a chirologist is only justified if he points out the course which leads to constructive and positive adjustment.

Thus a person who wants to follow the profession of a chirologist or graphologist should at least go through a psychological training or, better still, a psychological analysis, to make him conscious of the weaknesses and negative traits which otherwise might unconsciously influence his attitude. I know from my own experience that the widening of one's conscious sphere, which is the result of analytic work, also makes the chirologist a better teacher.

The basic principles of my chirological system, published for the first time in book form, are intended to instruct the chirological novice, or rather to support and supplement his instruction, for experience has taught me that chirological knowledge cannot be based on book learning only, but has to be acquired mainly by practical experience. In the following chapters I draw upon the experience of about thirty years and not on theoretical studies, and my observations and conclusions are based on numerous cases of all kinds and varieties, though most of them are based on people who are interested in psychology and have gone through inner difficulties and conflicts. As my work is chiefly empirical, my system is not a completely finished and rigid structure, but it is capable of expansion and adaptation and of assimilating new material which presents fresh angles of approach. This is the reason why I have not tried to superimpose my chirological teaching upon any extant psychological or philosophical system, though I gladly confess that C. G. Jung's school of thought and psychological method appeals most strongly to me and that I make use of his terminology. Other chirologists, however, have tried to graft their theory upon such systems but, according to my experience, it is impossible and futile to adapt the reality of the human hand which is the expression of life in all its facets to a more or less abstract and sometimes even irrational system of concepts.

In the subsequent chapters I shall expound the theoretical foundation of my system and conclude my book with practical examples based on the hands of children. The theoretical part, however, has a wider range than the practical section because the back of the hand, for instance, and the fingers of children under twelve are not sufficiently developed to allow of analysis. Another reason for the publication of this book on the hands of children is that I plan to follow it up with a work based on my considerable material of the hands of adults comprising normal, difficult and psycho-pathic cases. The present book is intended to supply the theoretical basis for its successor. The elaborate introduction is necessary because the people for whom this book is chiefly written—educationalists, teachers, priests and parents—would be unable to get a comprehensive idea from the prints of the hands of the children alone without the complete theoretical exposition of the hand of the grown-up. Some hands of children, however, from their earliest childhood already contain the complex maze of the lines of the adult and other signs, whilst in others only the four main lines appear.

This book is particularly important for parents because, as I have pointed out in the first chapter, the development of children is vitally conditioned by the conscious attitude of the parents, their unconscious and unsolved problems, and their false and misguided ambitions. If they become conscious of this they will develop into better models for their children and better educators, for, after all, the setting of a fine example has a greater educative force than active interference.

CHAPTER THREE

THE OUTER HAND

Once more I should like to stress that chirology cannot be learned from books only, but that practical experience and observation are as important as theoretical knowledge. It would be of little avail for the student to learn the subsequent details of my method by heart and to apply them as hard and fast rules ; he must assimilate and digest by constant practical application what he has learnt. One fundamental rule, however, should be always present in the mind of every chirological interpreter, whether he is a novice, apprentice or a master : on no account to judge an isolated feature and apply it generally to the personality as a whole but to interpret the feature in the light of all the other factors supplied by the outer hand, the fingers, the nails, the course of the lines, and so forth. Only the combined and interconnected features of the hand will yield a complete and reliable portrait of the personality. In a book which expounds a method it is, however, unavoidable to deal separately with each feature and its chirological significance, and the necessity of combinative adjustment can be pointed out only occasionally. Though this fundamental rule is universally accepted by all modern expressional systems, beginners as well as more advanced students are inclined to concentrate upon a striking or unusual feature and to make it the pivot of their final evaluation of the personality. Thus they unwittingly and involuntarily relapse into the old chiromantic practice of interpreting isolated signs, a practice which fundamentally differs from my method of appreciating and explaining the human character by viewing it in its entirety. All this and the effect of the previous generations upon the psychological make-up of a personality must be borne in mind by the student of psycho-chirology.

First of all the expressional tendency of the hand has to be ascertained. Just as our first impression of a person is of a general nature, taking in his face, his gait, his bearing, his movements and his voice, without paying attention to details, so we should let the outside of a hand affect us. This general expression (see fig. 1 lively, animated hand and fig. 2 rigid, inexpressive hand), of which the person is usually unconscious, conveys the spiritual quality, the inner standard, that is to say the nucleus of the personality. To grasp its deeper meaning we must be capable of direct contemplation, of intuitive perception, and of a complete absorption in the general aspect of the hand. I know from my teaching experience that this intuitive perception can be developed by the constant observation and study of hands. In the beginning all the pupils when asked to give an account of the general expression of a hand meet with failure and are convinced that they will never learn it. After much practice, however, they acquire the faculty. The initial difficulty arises because most people have lost or neglected their faculty of direct perception without thinking and without the rational recognition of details, a faculty which every naturally developed child possesses. The child, by virtue of an inborn collective knowledge, views things directly and in their totality. By collective knowledge we mean the sum of all experiences accumulated since the beginning of mankind which is present in every new-born human being, though unconscious and differently distributed and in a different degree. By a one-sided training of the intellect in controlled, definite and rational thinking, this collective knowledge is more and more forced into the background and in the end is lost altogether. Thus it is necessary to resurrect this submerged and inherent faculty of direct perception by systematic practice and training.

The expressional appearance of the hand shows the essential fundamental traits of the inner personality without, however, revealing the dominant, that is to say, that which is decisive in the dynamic development.

The outer hand reveals the static element of the personality, the inner hand its dynamics. By the static element I mean the consolidated intellectual-spiritual standard, the inherited disposition, the biological husk, as it were, which does not yet reveal to what extent these dispositions have either been active in the development of the personality or have remained untapped. Thus, for instance, the shape and appearance of the nails may reveal a disposition to lung diseases, yet the person need never be affected by it. An investigation of the case, however, would show that there has been lung trouble in the family which means that a certain amount of attention should be paid to this latent possibility in the children of this person. By the dynamic element which is represented by the inner hand I mean the effect of the milieu, the environment, the experiences and the unconscious upon the disposition and inherited traits as expressed by the outer hand. This yields the dynamic and, therefore, fluctuating portrait of the personality and its manifestations in the various periods of that person's life. Thus it must be always kept in mind that the outer hand does not allow of valid conclusions concerning experiences and development of the personality. The few exceptions in this respect are to be considered as anomalies such as pathological cases of people who are hereditarily doomed.[1] An exception of a different kind is that of a person whose hands and fingers are so strikingly different that they produce an impression of disharmony (see fig. 9). Here the divergency of the hereditary traits is so weighty that the dynamism and the range of experience are confined within narrow bounds. As the attitude to life of such a person is one-sided and lacking in elasticity, the impact of events is harder and has a more fateful effect than on a person who is less hereditarily conditioned and, therefore, more resilient.

As in the case of the fingers, so a striking difference in the expression and form of the right and left hands is of special chirological significance and must be carefully observed for the same reason. If both outer hands are essentially different we can assume that the character or the temperament of the parents must have greatly differed from each other. This divergency can affect the personality in various ways, for the diverse parental qualities may either assert themselves as conflicting forces and thus disturb and impede the person's development or they may be united and combined by a newly developed power, thus employing them in a positive and productive manner.

As regards the shape of the hand, we distinguish the three main shapes which are depicted in figs. 3–7, and two additional but much rarer types (see figs. 8 and 10). The three fundamental shapes hardly ever occur in their pure form, for as man is a mixture of a variety of qualities, so the forms of hands are mixed accordingly. The shape which is most commonly found in its more or less pure form is the rectangular hand. The student who wishes to ascertain the shape of a hand must first of all study the back of the hand and the fingers separately, for each individual finger may differ in form from the others as well as from the back of the hand. The more experienced chirologist, however, is able to take in the basic form of the

[1] The author was preparing a special publication on this subject.

hand as a whole, but when analysing the fingers he must establish the individual form of each one.

The conical hand (see figs. 3–4), which is also described as the aristocratic, æsthetic and beautiful hand, can be recognised by the fact that it is growing narrower towards the top. The fingers have a particularly pointed nail phalange. The conical hand is the hand of the mentally flexible, artistically receptive person who loves everything that is connected with art and delights in all that is beautiful. His enthusiasm is easily aroused, but he has little determination and consistency, is easily tired and finds it difficult to concentrate. Such people are inclined to go to extremes, and sympathy and antipathy are experienced at first sight. They also have a strong desire for variety, start all kinds of activities but very rarely complete what they set out to do. Should, however, the conical hand be supported by a very good and forceful thumb, the negative traits are counteracted and artistic results of a high order can be achieved. The emotional and intuitive elements predominate in this type of hand. People possessing such hands could be described as extravert in the Jungian sense, and the conical hand is indeed mainly found with people who follow an extraverted profession, such as actors, dancers and cabaret artists in particular. The ever-changing programme of their work and career amply satisfies their desire for variety.

The rectangular hand (see figs. 5–6), which is as wide at the bottom as it is at the top, is the opposite of the conical hand. Persons with rectangular hands are guided by reason ; they are practical, reliable, they think methodically, work systematically, but they are sober-minded and lacking in imagination, relying only on that which can be proved and practically realised. Everything that is new and does not come within the orbit of their rational experience is treated with scepticism and suspicion. Their personality is dominated more by intellectual than by emotional processes and, therefore they have little desire to delve into and to clarify their problems, but avoid these explorations of the self and pursue their established and accustomed course. In the symbolism of numbers the four represents earth, matter and surface. A rectangle with its opposite sides of equal length conveys the impression of something that is complete and limited, that is to say, not very capable of development, whereas a triangle appears to be much more mobile and progressive.

The spatular hand (see fig. 7), which is narrower at the base and wider at the top, is the hand of the active, creative person who has original ideas and the strong desire to communicate them to the outside world and yearns for sympathetic response to realise his aspirations. In contrast to the person whose hands are rectangular, that which is new and unaccustomed has a stimulating effect on him. In many books on hand-reading it is stated that the spatular hand is particularly characteristic of an egoistic type. This is a very superficial interpretation, for egoism can be found expressed in every form of hand. The truth is that people with decidedly spatular hands are often so full of or obsessed by their ideas that they work without consideration for others and themselves until they have carried them out. They are impelled by the wish to propagate their ideas and not driven by personal, self-seeking ambition. The spatular hand, however, is not the prerogative of intellectually and spiritually creative persons such as scientists, artists or doctors, but can also be found among working men and people generally who have independent, creative ideas.

The opposite type of the spatular hand is the rarely occurring philosophical hand (see fig. 8) which is characterised by inwardly inclined fingers and decidedly knotty joints of the upper and middle phalanges. The fingers are thin and bony. The philosophical hand indicates a mind with a strong inward tendency, keenly interested in and absorbed by religious and philosophical ideas and problems, the creative introvert who works, writes, paints or composes without desiring and needing the applause of the surrounding world ; on the contrary, he is rather shy and passive in matters of practical realities and longs for solitude and quiet.

The least differentiated type of hand is the primitive hand (see fig. 10), which has a very large, wide and bulky back, and stumpy, clumsy and inexpressive fingers. The thumb, particularly, is often so short that it hardly reaches the base of the index. The general impression of this hand is that of animality, undevelopment and clumsiness ; it is the hand of a person with a very limited scope for mental and spiritual development. Mainly peasants, and not working men, as is often wrongly assumed, have this type of hand. The fact that they are mentally and spiritually insensitive does not exclude achievements in the sphere of practical work.

The back of the hand is expressive of the practical intelligence and ability of a person and his attitude to the realities of life, to matter in general, to earthly things, mythically speaking the Earth, the Mother as symbolised by Mother Earth. As the back of the hand is limited on both sides by the knuckles and the wrist it is symbolically expressive of the finite, of limitation, of the unchanging element. Thus it represents the bodily sphere and the relationship to one's body.

The fingers, however, indicate the intellect, the spirit, imagination, and in a mythical sense Heaven, the Father as symbolised by the Father in Heaven, or the Heavenly Father. The fingers are symbolically expressive of the infinite—as they are unlimited and free at their ends—of that which is capable of development, of change within the personality.

Thus it is essential to observe the relative size of the back of the hand and of the fingers. If they are of relatively equal size a certain harmony and balance is indicated, but should the back of the hand be much larger and wider than the fingers (see fig. 6) we can infer that the practical, the material, the realities of life preponderate at the cost of the intellectual, the spiritual, the capacity for development and emotional sensibility. On the other hand, if the fingers are longer than the back of the hand (see fig. 5) the sense of reality is curtailed and the practical abilities are neglected to the advantage of feeling, imagination, intellect and spirituality which leads to conflicts with the world of realities and generates a tendency to live in a world of emotional phantasy and imagination.

Also the size of hand and body should be well proportioned, that is to say a tall person should not have too small a hand, and vice versa. In the first case the intensity and restraining power of the personality is not adequate, which may lead to a lack of self-control and restraint, restlessness and difficulty of concentration. In the case of a small person with too large hands we can assume that some quality of the personality is strongly exaggerated at the cost of some other quality, which produces greater inner tensions than the person can cope with.

People with large hands and long fingers are more emotional, more capable of enthusiasm and often more imaginative, and they are more inclined to obey their spontaneous impulses. Their hands are skilful and nimble, which makes them particularly suited for delicate work. In their work they pay special attention to detail and finesse, but not so much to the shape and arrangement of the whole.

Small hands, however, indicate people who are more sober-minded, less guided by their emotions than by reason. They think rationally, work and live with greater concentration, they are less devoted to the details than to the totality of a thing or matter. Thus they often neglect the detail in favour of the whole. Therefore the size of the hand and of the fingers often reveals how a person works, how he sets about it, and also shows the good points and weaknesses of his method of work. As a person with long fingers is better suited to minute and exact work which demands great concentration on detail, the person with small hands and fingers is more apt to do work which necessitates a more general approach.

Another essential factor in the chirological interpretation is the consistency of the hand, which can be ascertained by feeling the soft part of the angle between the base of the thumb and index as well as the part at the edge of the hand below the little finger. The hand must be relaxed and the thumb-index angle should be lightly touched with the thumb and the other spot should be simultaneously tested with the index (see fig. 11). The consistency is either soft, medium soft, medium hard or hard. The medium degrees of consistency need not be specially considered in an analysis. The soft consistency indicates an adaptable, understanding, flexible, mobile personality which on the other hand can be easily influenced, a person who is yielding, or even unstable and inconstant. The hard consistency shows a person who is difficult to influence, who is stubborn, tenacious, unyielding sometimes to the degree of rigidness and inflexibility. When investigating the consistency of a hand it is most essential to consider its shape at the same time. The conical shape, for instance, which suggests a flexible personality would be strengthened by a soft consistency, but such a high degree of flexibility may produce negative results such as instability of character.

The hard consistency, however, combined with the conical shape is an improvement because the flexibility is coupled with power of resistance and staying power.

The rectangular hand which suggests a sober, realistic personality is negatively enhanced by the hard consistency because the elements of rigidity, obduracy and immovability are increased, whereas the soft consistency adds movability and flexibility to the qualities of the rectangular hand, and thus improves it.

People with hard hands are mostly inclined to exaggerate, to be troubled, to acquire complexes and fixations caused by their tendency to shut themselves off, to harden themselves. Those who have hands of a soft consistency are more easily influenced by their environment and the outside world generally.

The normal colour of the hand is pink with a greyish tinge. The hands of excitable, violent, very active and sometimes brutal people are too red. Hands of pronounced whiteness suggest leisurely, passive, lazy and sometimes insincere people, particularly so when the whiteness is combined with a soft consistency. The colour of the hand, of course, depends upon and is connected with the circulation of the blood, for too weak a circulation produces a white and too strong a circulation a too red colour. The interconnectedness of psychical and physical phenomena which manifests itself here will be discussed at greater length in the course of this book.

In conclusion we must touch upon the formation of the so-called knots (see fig. 12), excrescences which can appear at the upper, middle and lower joints of the fingers. They signify an enhancing of the qualities concerned whether they are positive or negative. They hold or grip fast just as a knot holds, grips, fastens. In the case of gout, for instance, the body holds something which is an impediment, an encumbrance. In a positive sense knots suggest greater staying power and support, and apply to material things when they appear at the lower joint, to the mental capacity at the middle, and to imagination and feeling at the upper joints. If they are too strongly emphasised the quality in question is overstressed. A conical hand is improved by knots because they give stability, whereas knots at the lower joint emphasise the orderliness suggested by a rectangular hand to the degree of pedantry.

CHAPTER FOUR

THE FINGERS

The fact that body and soul are an indivisible whole is generally accepted by modern psychology and psychiatry and, therefore, need not be expounded in this book. The psycho-chirologist is, however, greatly concerned with the interconnectedness of organic and psychic diseases, for many illnesses of the body or the mind cannot be properly understood and explained without an investigation of both aspects. This interconnectedness enables the chirologist to discover certain dispositions to illnesses and diseases, though I should like to stress that it is not within his province to diagnose illnesses or to give medical advice. He can, however, be of assistance to the doctor by suggesting to the person in whose hand he has found a disposition to a certain illness, that he should seek medical advice, a step which may prove prophylactic. A word of caution must be added in this respect, for most people who consult a chirologist are more or less open to suggestion and particularly inclined to be influenced in matters concerning their state of health. As stated above, various features of the hand sometimes provide us with the possibility of discovering in time certain dangers to a person's state of health and of ascertaining with certainty which organ or organs of the body are weaker than the rest. The fingers, which are expressive of the physical and psychical functions of the personality, are very revealing in this respect, but before dealing with this physiological aspect we shall devote our attention to their general chirological significance.

Among the fingers, the thumb is the representative of vitality, of the life-force, of physical power and thus of self-assertion and will-power. Its shape, expression, arrangement, position, relative size, are very essential factors for the estimation of the forces of the personality which are concerned with an individual's assertion in and mastery of the outside world. Four main features have to be observed in connection with the thumb : its form, its position in relation to the rest of the hand, its articulation, that is to say the way in which its two phalanges are arranged, and its degree of flexibility. The shape of the thumb discloses a person's vitality as such, but not the amount of energy actually displayed or the degree of self-assertion. Thus a long powerful thumb indicates a person full of vitality, and an according display of vivacity, manifold interests, a great receptive capacity but not the degree of energy employed. This is expressed by the upper or nail-phalange.

By way of articulation the thumb is divided into a lower and an upper phalange. The two phalanges should be well proportioned, that is to say, of more or less equal length and forcefulness if the qualities they express are harmoniously distributed.

The lower phalange tells us to what degree reason and reflection control the actions and manifestations of energy of a person. If the lower phalange is as long and as broad as the nail-phalange, reasoning power as well as will-power are well balanced. Should, however, the lower phalange be very narrow and the nail-phalange very broad, producing a waistlike shape in the lower part of the thumb (see fig. 13), the person is bound to act without much deliberation, that is to say, impulsively. If in such a case the nail-phalange ends broadly, we know that the will-power, energy and staying power of the person are good, but guided by spontaneous and impulsive forces and not by reason. When the lower phalange is very narrow and the end of the nail-phalange is very narrow too, the person is impulsive and has a weak will. If the lower phalange produces the above described waistlike line, but is forceful and longer than the upper phalange, we can conclude that the person acts impulsively and reflects afterwards. The length of the lower phalange enables him, however, rationally to advise others as to what course they should take in their actions.

A large and broad lower, combined with a narrowing upper phalange (see fig. 14), is a sign that reason prevails over will-power, that is to say, the person reasons and reflects to an extent which kills action. Whereas a person with a narrow lower phalange acts first and deliberates afterwards, which is characteristic of the extravert as described by C. G. Jung, the person with a broader lower phalange is a more introverted type who thinks first and acts afterwards, or does not act at all. If both phalanges are of equal length and breadth, as already stated, the powers of will and reason complement each other harmoniously.

A person who has a very long thumb and a weak or narrowing nail-phalange possesses vitality but little staying power when it comes to the execution of his intentions. Such people are usually vivacious and have varied interests, but show a strong tendency to fritter away their energies, and find it difficult to concentrate. This discrepancy of vitality and will-power sometimes produces criminal tendencies because the needs of such a person's vitality cannot be satisfied by a normal expenditure of energy. If, however, a thumb is relatively small as compared to the hand as a whole, but has a broad and powerful upper phalange, we conclude that in spite of a poor vitality, and often physical weakness, the person possesses strong psychical energies. Such people are capable of great feats of energy, but in order to make up for their physical deficiencies they are inclined to overstrain their capacity. They should be advised not to subject their dynamic energies to too severe a strain but to interrupt their work by regular intervals of relaxation.

An important factor for the estimation of the display of energy is also the way in which the lower and upper phalanges are linked. One can often observe that the part where they link up, the joint which connects the two phalanges, shows smaller or larger hump-like elevations (see fig. 15). They are an indication that the energies are unevenly applied. People who have such humps on the middle joint of the thumb tackle things with great energy and, in the beginning, even with extraordinary verve, but in the course of the execution of the task their efforts flag. The psychological reason for this behaviour is an uncertainty, an unconscious fear of failing in their attempted purpose, which gives all their activities demanding an expenditure of energy a certain restlessness and unevenness. As the thumb, being the finger of life, represents the spring of vitality, it is not merely the dynamic expression of a person's display of energy but of the whole rhythm of his life. Thus it is frequently observed that the various periods in the life of people whose thumbs have these hump-like protuberances are full of ups and downs.

A feature which has to be considered carefully is the relationship of the thumb to the four fingers as well as the angle it forms with the hand, that is the distance between thumb and index. When ascertaining this angle the hand must always be in a relaxed position. An angle of approximately 45 to 60 degrees (see fig. 16), which can be regarded as normal, indicates a self-reliant person, a person who wishes to be independent, whose attitude to life is unconventional, who desires a certain measure of freedom in the conduct of his life. These qualities, if supported by the general appearance of the hand, are often combined with generosity, frankness, and an unbiased approach. If the angle is less than 45 degrees (see fig. 17), which means that the thumb is approaching the hand very closely, we conclude that such persons are more dependent on conventions and the opinions of their environment, less independent in their actions and more or less unfree in their general attitude to life. This does not mean, however, that such people must remain so for ever because the upper phalange which expresses will-power, and the forcefulness of the thumb as a whole, must be considered as well. If the thumb is powerful and the upper phalange well developed, their owner's deficiency may be due to a psychological disturbance which could be eliminated by psycho-therapeutic treatment. Should, however, the shape of the thumb and the upper phalange confirm the weakness, we may conclude that the person has a disposition for a certain dependence on conventions and is lacking in self-reliance. If the angle of the thumb is more than about 60 degrees (see fig. 18), branching off in an almost horizontal line from the hand, the independence and self-sufficiency of such persons is so extreme that it becomes identical with an absence of adaptability and expressive of an asocial attitude. These people are without consideration for others and usually also tactless and devoid of inner refinement. This characteristic position of the thumb is often found in classes of society whose upbringing and education is a flagrant contradiction of their inner attitude.

If the upper phalange is not the straight continuation of the lower but has an inward bend which gives the thumb a broken appearance (see fig. 19), we conclude that the self-confidence of the person is somewhat disturbed and very easily upset, or that he is suffering from inferiority feelings. The broken appearance of the upper phalange which represents will-power, chirologically indicates that the development of the will-power is impeded. This disturbance and inhibition of the forces of the will necessarily have a blighting effect on the development of a self-reliant personality.

Finally we must consider the suppleness, or stiffness, of the upper phalange (see fig. 20). A supple thumb, whose upper phalange has a slight outward bend, suggests flexibility, suppleness, the capacity of appreciating the feelings of others, vivid emotions but also greater instability, susceptibility to influence from outside and a tendency to give in.

A stiff thumb (see fig. 21) indicates obstinacy, tenacity, an unyielding mind to the point of rigidity, pigheadedness, and little flexibility.

If the upper phalange is so flexible and bent back that it almost touches the lower phalange (see fig. 22), we are dealing with a person who can penetrate a matter or personality to the point of identification. People of that type may read a book or listen to a lecture and assimilate what they have read or heard so strongly and intensely that it becomes their own. They are convinced that they were the originators of the ideas put forward and do not hesitate to say so. This feature combined with other traits also applies to the material sphere and people thus afflicted are unable to appreciate the difference between their own property and that of others, in other words they are kleptomaniacs, that is to say, thieves without economic necessity.

An abnormal thumb is the so-called club-shaped thumb (see fig. 23) with a bulbous upper phalange which gives it a clumsy and crude appearance. In it the energies are pressed into an uncontrolled, formless mass, and like all unformed and overstressed things manifest themselves in an explosive manner. The club-shaped thumb suggests people of uncontrollable temper, persons who in their fits of blind rage may commit acts of brutality or even deeds of violence. It is often called the thumb of the criminal or killer because it is an almost unvarying characteristic of the hands of criminals of a violent or homicidal kind.

Sometimes, however, the club-shaped thumb deforms the hands of girls and women who neither fly into ungovernable passions nor delight in savage brutality but on the contrary, seem emotionally and temperamentally extremely inhibited. A closer investigation of these cases brought to light that at least one of the ancestors of these women had in a fit of rage perpetrated acts of violence or committed manslaughter. The dreams of these women betrayed that in complete contrast to their inhibited behaviour deeds of violence played a prominent part in their unconscious. Their inhibition was the result of the discrepancy between the conscious and the unconscious layers of the mind.

A fairly frequent feature is a thumb which starts deep in the Mount of Venus (see fig. 24) and thus allows only a small angle to be formed by thumb and index. This thumb which does not rise in a free movement from the Mount of Venus—the Mount of impulses and instincts—but is closely fettered to it tells the chiro-analyst that part of the libido is strongly bound up with the father or the mother constituting a parent-fixation which holds part of the powers of self-assertion and will in infantile bondage. To compensate for these arrested energies, these people cultivate an exaggerated ambitiousness in a domain outside their sphere of work such as sport, or they become inveterate collectors. Their ambitiousness is often combined with an almost childish and obstinate pride which does not permit them to accept other people's advice or guidance because they must always have the feeling that they owe everything to themselves. This over-sensitiveness is caused by the lack of self-reliance which was produced by their parent-fixation.

The appearance of the nail of the thumb shows whether the composition of the blood is normal. If the thumb nail alone is of a pale or bluish colour the composition of the blood is not satisfactory. Whether this is due to a lack of phosphorus, calcium or iron, or whether the distribution of red and white corpuscles is defective can only be ascertained by a medical blood test.

The Index

Like the thumb, the index is connected with the outside world, but whereas the thumb represents the powers and energies which are directed to and concentrated on the outward life seeking to master and firmly to hold life and its phenomena, the index actually points at the outward phenomena. Whereas the thumb of all fingers is the one which seizes objects and holds them fast, and possesses the greatest strength for doing so, the function of the index is almost exclusively that of touching and of pointing out the phenomena of the external world. I have, therefore, called it the "World-finger," that is to say, the finger which is most actively connected with matters of the surrounding world. The child in Continental schools when told to raise his hand if he wants to catch the eye of the teacher automatically lifts up the index, and all European peoples use their index if they wish to indicate something within their range of vision. Just as we observed four main features in connection with the thumb, we similarly connect four features with the index : power of observation, sense-perception, adaptability, and the urge of self-assertion.

The sign of well-developed powers of observation directed upon the outside world is a straight index, whereas a bend sideways indicates that they are inadequate (see fig. 25). The lower part of the index is concerned with the observation of the nearer environment, the upper with that which is more remote, as seen for example in a sense of direction or the ability to find one's bearings in town and countryside. A good observer is a person who is able quickly to take in the phenomena of the outside world in contrast to the intuitive person who has a quick grasp of the phenomena of the inner world. If a person immediately notices that a musical instrument or some other object which is usually in a room has been removed, he does so on account of his powers of observation. Should he, however, find that the atmosphere of the room has changed without noticing the absence of the object in question, he has been guided by his intuitive powers.

The sense-perceptions which are indicated by the index are connected with the sense of smell, taste, touch and—as far as colour is concerned—of sight. An index of a particularly animate and differentiated expression and of conical form (see fig. 26) suggests a refinement and heightened sensitivity regarding these particular perceptions. An index which also indicates the degree of adaptability to the circumstances of the outside world should altogether have a conical form, because as pointed out in the preceding chapter, people whose hands are conical adapt themselves more easily. If the index is rectangular, the adaptability of the person is poor. The same holds true if the distance between index and middle finger of the relaxed hand differs from the distances between the other fingers (see fig. 26), which makes the index appear isolated, or if it is noticeably bent towards the inner hand or has an outward bend.

Adaptability is the ability to conform to the necessities and facts of the outside world in an inwardly harmonious manner, that is to say, to accept these given facts and circumstances after their conscious investigation and appreciation. This does not mean that one should uncritically and indiscriminately bow down to these given facts and necessities but on the contrary one should probe and try to attune them to one's innermost being, even if this demands a sacrifice from the individual ego.

Laotse, the great thinker and sage of ancient China, said about adaptation :

The right life resembles the water which adaptable to everything adapts itself to everything ;
The farther away it leads from the workaday things
The closer it approaches the path.

Thus :

It is earthliness in the realm of existence,
depth in the realm of the soul,
love in the realm of emotion,
sincerity in the realm of thought,
development in the realm of purpose,
power in the realm of activity,
appropriateness in the realm of the deed.

Fundamentally :

Adaptation eliminates suffering.[1]

What the ancient sage expressed in his inimitable way C. G. Jung tries to formulate in a more Western and modern fashion when he talks of the sociality of the individual which does not mean the giving up of individuality but " more intensive and more general collectivist coherency " (C. G. Jung, *Psychologische Typen*, p. 637, [*Psychological Types*]). Experience has taught me to estimate the sociality of a person by observing the relationship of the world finger to the other fingers.

Thus if the index stands too far apart from the middle finger without a very considerable inward or outward bend, the adaptability is poor because the person is not sufficiently inclined to take an interest in the manifestations of the outside world. Should the finger be too far inclined to the outside (see fig. 27), the adaptability is also poor because the outside world exerts too powerful an influence or the person is too dependent upon the recognition of his personality by his environment, a position which is overcompensated for by an exaggerated sensitiveness and too much concern for the environment at the cost of the inner life. Should the index show a too pronounced inward bend (see fig. 28), we conclude that the person devotes too little attention to the outside world because he is completely engrossed in himself, which also leads to hyper-sensitiveness or a fear of the realities of life. Persons of poor adaptability are

[1] *Die Bahn und der rechte Weg des Lao Tse*, Der chinesischen Urschrift nachgedacht von Alexander Ular (Insel Verlag, Leipzig 1912), p. 14.

mostly very good companions at parties and, apparently, adapt themselves well to society. The explanation of this paradoxical fact lies in their inner unconcern. It is much easier for them to adapt themselves outwardly and superficially because inwardly they remain reserved.

The urge of self-assertion can be gauged by the length of the index. If it is longer than the third finger we conclude that the person wants to play an important part, tries to dominate and to rule others. Should a long index be supported by a good thumb or a forceful upper phalange of the thumb, which is expressive of will-power, the person is in a position actually to realise his desire of self-assertion. If, however, the long index is not backed up in this way, the will-power which is needed to translate the urge into a reality is lacking, and the discrepancy between the lust for domination and ineffective self-assertion consequently produces inner conflicts and neurotic symptoms such as stammering, claustrophobia, and other physiological phenomena brought about by psychical causes. Thus the discrepancy of a too long index and a weak thumb is a very essential factor in the search for the explanation of many a neurotic enigma.

It is an interesting fact that almost all schoolmistresses whose hands I have seen had these slightly too long indexes, a phenomenon which could not be observed among the same number of male teachers. The explanation of this is, I suppose, to be found in the fact that women have a natural inclination to take up teaching, an inclination which is fostered by an unconscious desire to guide, lead and dominate others. Men often become teachers not because they are particularly keen but because it is the wish of their parents. It would, however, be wrong to think that a long index is an exclusively negative feature, for it can also denote, especially when well shaped, positive qualities, as for instance an ability to guide, to lead and, provided the index has a certain form, to organise. The necessary condition, however, is that the thumb is strong and its upper phalange good.

If the index is smaller than the third finger the person is not interested in matters concerning the outside world, and meets with difficulties when he has to carry out practical plans or projects because of his poor social adaptability to his environment. It is, of course, essential that the general expression of the hand should be considered when analysing this or any finger. Thus the realistic or materialistic expression of the hand is enhanced by an index with an outward bend. Such a person is inclined to overestimate everything connected with the so-called realities of life and thus is subject to influences from outside. If, however, the general expression of the hand shows an inward trend of mind, an interest in spiritual things, the same position of the index signifies that the outside world is apt to have a disturbing and strongly irritating effect upon the person, who, therefore, is inclined to avoid and shun tasks which confront him in the sphere of reality.

Thus the above-mentioned functions of index and thumb which are so closely bound up with the environment belong to the sphere of consciousness. The organs whose sensitiveness, vulnerability and disposition to illness can be gauged by the appearance of the index are also more or less connected with the outside world and conscious experience. These organs are the stomach, the liver, the gall and the spleen. If the index is bent sideways towards the middle finger (see fig. 25) and has a strikingly pale-coloured nail we may assume that one of the enumerated organs is delicate and susceptible to illness. Though these organs of the human body are supplied and nourished from the outside, it is possible to ease, if not to cure, their illnesses by a correct attitude of the mind which, by the way, can also produce a prophylactic effect. This preventive effect can be obtained by adjusting one's diet in accordance with the susceptibility of the particular organ, or in case the illness has already set in, by strictly obeying the orders of the doctor regarding diet and conduct. This seems to be a superfluous thing to say but experience shows that a great number of patients think differently. Also another phenomenon which is connected with the correct or wrong attitude in the sphere of reality can be studied at the index: a disposition to rheumatism, gout or sciatica. This is expressed by a very prominent thickening of the lowest phalange of the index, forming a striking contrast to the other phalanges of the index and to the corresponding phalanges of the other fingers (see fig. 29). Quite frequently the cause of this pathological disposition can be attributed to a too one-sided diet and a predilection for alcoholic drinks, but this disposition can be counteracted and illnesses can be prevented by a correct attitude, that is to say by a moderate and adequate diet.

For reasons which will become apparent later, the discussion of the middle finger will conclude this chapter, and we therefore turn to :

The Ring Finger

Whereas thumb and index are connected with the environment and the sphere of consciousness, the ring finger displays the function of the emotions bound up with the inner life of the personality, and it shows the creative power of the mind, the intuition, and artistic talents and interests. Thus the ring finger is essentially a finger expressive of unconscious functions. With the ring finger not only the form but also the degree to which it is animate is of greatest importance, though I must admit that it is difficult to convey in words what I understand by " animateness ". The range of the finger comprises the emotional domain as related to the entire environment, with the exception of the partner-relationship.

Should the ring finger be straight and rectangular we assume that the emotional side is in harmony with the rest of the personality, particularly if it equals the length of the index. In case it is much longer than the index an over-emphasis of the emotions is indicated, a too exclusive preoccupation with one's own world and a lack of appreciation of the values of the outside world of realities. This may lead to a too strongly stressed introversion of the emotions, particularly if the long ring finger has a decided inward bend towards the palm. Also the popular denomination of " ring finger " confirms that the third finger is connected with the emotions, for on the whole the ring which is worn as the symbol of an emotional tie with another person is placed on that finger. I have made the observation that people who wear this ring on the first and not on the third finger are emotionally cold and have a great desire to attract attention. A bend sideways of the ring finger to the second finger (see fig. 1) conveys that the emotional function is undeveloped and still quite unconscious. This crooked ring finger is usually found with people whose life seems to be an uninterrupted chain of disappointments, who inevitably find themselves in situations which lead to conflicts and difficulties in their own self as well as with others. It is interesting to note that such people are constantly disappointed and disillusioned because they almost invariably approach others with wrong expectations and ill-applied demands, and instead of feeling relieved that they only deceived themselves are resolved to be disappointed, thus preferring a negative and harmful result to a positive experience.

The range of psychical development and of the intuitive powers of a person can be ascertained by the more or less animate appearance and the particular expressional quality of the ring finger. If in comparison with the other fingers the ring finger is particularly animate, expressive and differentiated the creative power of the psyche is particularly well developed. The intuitive capacity can be observed by the same signs, but the effectiveness and application of the intuition within the mind as a whole can only be judged by the interpretation of the Lines of Intuition. Though the discussion of the various Lines of Intuition is reserved for a later chapter, I should like to explain now what I mean by intuition.

Fundamentally, intuition is an unconscious force, the inborn creative force which every normal human being possesses, which independent of conscious experience enables him to understand the inner being and the inner connectedness of things and persons. In its substance it is not subject to the influence of or affected by the conscious mind but draws upon the unconscious. In contrast to the qualities expressed by the index, which are connected with observation and sensation as far as the external world is concerned, intuition is focused on the inner phenomena. With every normal child the intuition is the most powerful and most active force. Its effectiveness, however, is endangered by a too rigidly conventional education or a one-sidedly intellectual training at school. Unfortunately most parents model their educational plans according to a stereotyped pattern, and allow themselves to be guided by their own, mostly inadequate, experiences instead of considering the nature and character of their child, thus doing harm to the original intuitive ability and reducing its value and effectiveness for the personality. The damage done to the intuitive powers often leads to a lack of confidence in their reliability and is chirologically expressed by a strong forward bend of the ring finger towards the palm and its inclination towards the middle finger (see fig. 30). The suppressed or inhibited intuitive power can, however, be effectively restored by making the person conscious of the position or by eliminating the inhibiting factors. Artistic talent is expressed by a straight, expressive and spatulate ring finger. The particular nature of the artistic talent can only be ascertained by a combined interpretation of various features of the inner hand such as mounts, lines and others.

Whereas the index reveals the inferiority of organs which are connected with the conscious mind and the outside world the ring finger represents the heart, an organ which is not influenced by the conscious mind and not sustained from outside but from within. Consciousness can do very little in the case of heart diseases. Once more we have found a connection between psychical and physiological functions. A disposition to nervous or organic cardiac diseases can be diagnosed if the ring finger is crooked and the Moons—which will be discussed later on—are missing on all the nails.

The Little Finger

The fourth finger reveals that part of the emotions which is exclusively connected with the partner-relationship, and shows the manifestations of that part of the libido which is bound up with the sexual life. As far as its position and relationship to the other fingers (see fig. 30) are concerned it is usually somewhat concealed or covered by the other fingers because it expresses the function which is a most personal and intimate feature in the life of man. The thumb which indicates the life-force as such and the little finger which expresses the sexual manifestation of this force occupy the two extreme positions in the hand. With the little finger too, four things have to be considered: its form, expression, shape and relative size. Apart from the sexual function it indicates musicality, a gift for speech and languages, vocal ability, sense of form and a gift for oratory. The fact that talents connected with language and voice are expressed by the very finger which indicates the sexual function is not really surprising if we consider that the pitch of a eunuch's voice changes after castration, or the well-known effect puberty has on the voice of a boy. And last but not least we all know from our own experience to what an extent the speaking or singing voice influences the mood and is able to create an atmosphere of pleasure, joy, delight or rapture, states of mind which have a more or less erotic quality.

A little finger of normal length, that is to say, extending to the base of the third phalange of the third finger, and of clearly conical shape, supported by a flexible thumb and an adequate third finger, indicates musicality. If the tip of the finger is slightly rectangular (see fig. 32) the person is phonetically linguistically talented. A little finger of more than normal length or one which even reaches the beginning of the nail part of the third finger suggests a good negotiator. Persons who have this kind of little finger can very cleverly exploit the weaknesses of their opponents by letting them do most of the talking, and then launching their attack suddenly they take the opposing party by surprise. If the little finger is straight and completely rectangular (see fig. 33) the person has a pronounced sense of form, a talent which according to the type of the hand finds expression in the literary field or in speaking, or is applied to architecture or composition. Should the rectangular-tipped little finger be particularly long the sense of form is overdeveloped and the desire of formulating too exaggerated. In such cases the preoccupation with the form is so strong, sometimes to the degree of obsession, that often the subject-matter is neglected or lost sight of. A grammatical-linguistic gift is chirologically expressed by a small rectangular little finger of less than average length in a rectangular hand.

To gauge the sexual function correctly we must consider the Mount of Venus as well. The relationship and position of the little finger to the hand as a whole, however, strikingly shows the adjustment of the sexual function within the psyche as a whole. If the little finger is clearly divorced from the other fingers (see fig. 31) we may conclude that the sexual function does not fit harmoniously into the entire personality. This may be due to an under- or over-emphasis of the sexual, that is to say, the person is inhibited, disturbed or easily irritated in this respect or he is without inhibitions, physically insatiable, and solely interested in sensual pleasures. In either case be it by a too strong or too feeble emphasis of the sexual within the personality as a whole, the sexual is set apart, singled out and, therefore, plays relatively too big a rôle in it.

This characteristic position of the little finger can be strikingly illustrated by two parallels from the history of art and from the history of civilisation. The former shows that on almost all old pictures of the Madonna her little finger is striving away from the others, an interesting feature which was once explained to me by a writer on art who told me that probably the sexual note of the models used by the painters was either too strongly emphasised or too feebly developed. The second illustration is taken from the history of civilisation. At the end of the nineteenth century in the period of the emancipation of women, when equal rights were also demanded in sexual matters, it was considered as particularly fashionable and distinguished to hold the teacup with the little finger daintily jutting out, a symbolic gesture which manifested this particular demand very clearly.

The jutting-out little finger combined with certain features of the thumb sometimes reveal a strong desire for independence regarding the partner, which is often coupled with an urge to isolation and a fear of responsibility. This position can also signify that the person is able to maintain a partner-relationship on an intellectual-spiritual basis without physical relations, or vice versa.

There is a correspondence between the little finger and the index, for both are often standing apart from the other fingers, a parallelism which tallies with the psychological observation that a difficulty in the adaptation to the environment almost invariably implies a difficulty in the partner-relationship.

All these phenomena must, of course, be considered in connection with the position and expression of the hand as a whole and with the Mount of Venus. If the little finger is not jutting out but clings to the hand (see fig. 30), that is to say is closely attached to the ring finger, we conclude that the person has a great longing for tenderness and affection, a disposition to cling to others combined with shyness and bashfulness. A little finger bent towards the palm and clinging to the ring finger (see fig. 34) indicates prudishness.

These manifestations of the jutting-out little finger indicate difficulties in the partner-relationship. A further deviation from the normal position is the deeply rooted or deeply set-in little finger (see fig. 35) indicating a disposition to difficulties in the partner-relationship. People with this kind of little finger have a father or mother fixation and, therefore, in their phantasy worship the ideal image of the parent concerned, an *imago* which in most cases unconsciously controls and influences the partner-relationship. Unconsciously all partners are compared with this unattainable ideal image, an attitude which either results in a series of disappointments in their relationships with the opposite sex or in difficulties of sexual adaptability. If this position is strengthened by a deeply set-in thumb, the difficulties of adaptation are considerably increased because of an additional tendency to dominate the partner.

The organs connected with the little finger belong to the abdominal region : they are the uterus, the ovaries, the testicles, the bladder and the kidneys.

Uterine illnesses such as an indentation or dislocation of the uterus can be recognised by an indented second phalange (see fig. 36). A particularly thick lower phalange indicates that the ovaries are affected. Bladder and kidney trouble are revealed by an inward bend of the top phalange, and by the fact that it is the only finger without a nail moon.

The Middle Finger

We have seen that thumb and index are bound up with the sphere of consciousness and ring finger and little finger with that of the unconscious. The middle finger, however, assumes a central position and has, as its name suggests, the function of a mediator. Apart from that, it indicates the intellectual function, for, after all, it has to mediate between the conscious and the unconscious spheres, those domains of the personality which are concerned with the outside world and the inner world of the self. Thus it has to be a mediator between the *ego* and the *id*. As a mediator must be able to take a view above the contesting parties, it follows that the middle finger should be taller than all the other fingers, though it should not surpass the second longest finger by more than a nail's length, for if it goes beyond this limit it forgoes its rôle as a mediator and usurps independence (see fig. 37). This would mean that the intellectual function surpasses all other functions, that the person thinks too much about himself, that he isolates himself, or in other words that he is too exclusively directed towards his own self. This overstressing of the intellectual side upsets the harmony and balance of all other functions, a state which often leads to fits of depression. If the middle finger is too short (see fig. 38), let us say not longer than the normally second longest finger, the person does not think enough and, therefore, is inclined to act impulsively and rashly or to lead a happy-go-lucky or even reckless life. The short middle finger often occurs among artists, who live and work more intuitively. A completely rectangular middle finger of normal length shows a methodic and logical mind. If the finger is crooked and not animate in expression, the intellectual capacity is undeveloped and unconscious. We see it is the length of the middle finger which is related to the position of the intellectual function within the personality, and the form of the finger which pertains to the intellectual function as such.

Once more we find the connection between psychical and physical functions, for the middle finger indicates a pathological disposition or sensitiveness of the intestines. The intestines are a mediating organism which can be influenced and affected both consciously and unconsciously. If the top phalange of the middle finger is bent to the side of the third finger and the lowest phalange is particularly thickened and a certain line appears in the palm which will be described later on, we can conclude that there is a disposition to enteric illnesses.

In order to estimate the relation of the various functions to each other it is essential to ascertain the relation of the fingers to each other and their relation to the middle finger. This particular point, as well as the peculiarities of the individual fingers, will be dealt with in the following chapter.

CHAPTER FIVE

THE POSITION OF THE HAND
THE RELATIONSHIP OF THE FINGERS TOWARDS EACH OTHER
THE PECULIARITIES OF EACH FINGER

The preceding chapters have made it clear how important it is to study the relative position of the fingers in order to judge the various functions and their interconnectedness. Two important things have to be distinguished in this respect : the hand raised in a relaxed position with its back facing the observer, (see fig. 39) and the relaxed dropping of the hands with their palms lying on the table (see fig. 40). The raised position reveals the conscious attitude and position of a person towards the environment, which according to Jung's terminology would be the position of the *persona* within the entire psyche. The relaxed dropping of the hands conveys the unconscious inner attitude towards the realities of the environment. The result of the dropping of the hands can be considered fairly harmonious if the palms touch the supporting surface with the Mounts of Venus and of the Moon (see fig. 40). In both these procedures it is important to observe the position of left and right hands towards each other. If the uplifted hands are pronouncedly inclined towards each other, we conclude that the person is dependent upon his milieu, that he has a " milieu-fixation " (see fig. 41), an attachment whose chirological significance will be discussed at greater detail in the chapter about the lines of the hand. The more the hands are inclined towards each other the stronger the milieu-fixation. If the left hand assumes an upright position and the right hand is strongly inclined towards the left, we can assume that the fixation has to be attributed more to the milieu than to the person. Reversely, if the right hand is in an upright position and the left is strongly inclined towards it, the fixation is mainly conditioned by the person himself.

When the hands, after their relaxed dropping on the table, do not lie in a parallel position but are clearly inclined towards each other (see fig. 42) we conclude that the milieu-fixation is the central problem and strongly preoccupies and influences the person. In the case of a fairly young girl, for instance, I found that the right hand had dropped far away from the left but was pointing towards it. A further investigation showed that the milieu-fixation had been caused by the girl's strong opposition to her mother who had neglected her. In protest the girl had tried to break away from her milieu and later the mother had endeavoured to hold her. In connection with the relaxed dropping of the arms, attention should be paid to the manner in which the inner surface of the hand and the fingers touch the table and also which of the fingers and which part of the hand do not. The best results are obtained on a moderately high table without a cloth. If the hand touches the surface evenly (see fig. 40), we assume that there is a certain harmony between the attitude of the person to his environment and his inner life. When the hand or the fingers do not touch (see fig. 42), this harmony does not exist, but there is a discrepancy in the inner and external attitude of that part of the psyche which is represented by the particular part of the hand, often manifested by an irrational attitude or escape from reality into a strongly egocentric inner world. The experienced chirologist will also be able to infer whether the person is inclined to be an extravert or an introvert, for extraversion will be revealed either by a slight, or no, inward curving of the fingers, introversion, however, by a pronounced curving (see fig. 42). If the position of the uplifted hands, or of the fingers towards each other, greatly differs from that produced by the relaxed dropping of the arms, we conclude that there is a considerable divergency between the conscious attitude of the person to his environment and his inner attitude.

When dealing with the uplifted position of the hands we must also observe how their backs join the fingers. If the joining process is harmonious and smooth (see fig. 43) without the knuckles jutting out prominently, our assumption is that the attitude of the person to the tasks and problems of practical life and the intellectual and spiritual tasks of the personality is one of harmony. Should, however, the linking up be impeded by very prominent lower phalange joints (see fig. 44) producing the effect of a separating bulwark between the fingers and the back of the hand, the attitude to the given realities of life and the aspirations of the mind is disrupted. An important factor in this connection is the relationship between the back of the hand and the fingers, that is to say whether the proportion of the back of the hand as compared with that of the fingers is fairly good, a feature already mentioned in Chapter Three. If, for instance, the back of the hand is strong and large and the fingers correspondingly long but the joining up bad, that is to say not in a smoothly flowing line, it may be that the intellectual and spiritual element is over-emphasized and the values and necessities of practical, concrete reality are underestimated, or sometimes even treated with contempt. This does not mean that the practical ability of the person is inferior, for as a rule this contempt for the practical side of life is mostly the product of an opposition to a too strongly emphasised materialist tendency of the milieu. If, however, the back of the hand is small and the fingers much longer and stronger but badly linked up with it, the discrepancy arises because the person feels much safer and more at home in the intellectual and spiritual than in the practical sphere. There is no opposition at the bottom of such a person's attitude ; he is simply unable to cope with the problems of practical life. The discrepancy between greater intellectual assuredness and helplessness in matters of workaday life sometimes produces neurotic phenomena such as stuttering, claustrophobia and fear of life.

Another feature which has to be looked out for is the direction in which the fingers are inclined, whether they bend outward or inward towards the palm of the hand. The outward direction (see fig. 45) suggests a person who is optimistic, easy-going, sometimes even irresponsible and reckless, a person who easily establishes contact with others and on the whole is more open to the demands and temptations of the outside world. People whose fingers have an inward direction (see fig. 46) are very interested and wrapped up in their own selves, they are very egocentric, sluggish and sometimes dejected. A superficial observer might be tempted to label the first type extravert and the second introvert. The chirologist must, however, proceed here very cautiously, for it may well be, and is very often the case, that these attitudes are not genuine but the outcome of an inhibition or psychological distortion, and consequently no valid conclusion can be arrived at as to the original disposition of the person. To interpret the position of the hand correctly it is most important to establish which fingers are concerned and how they are placed towards each other. It sometimes happens that not only the position but even the shape of the fingers is changing according to the development of the person. All fingers can change their position, but so far I have only been able to observe the changing of the shape of thumb and index.

Once I experienced the metamorphosis of an index within the short period of three months. That particular index of the left hand was crooked (see Chapter Four), a sign of poor powers of observation and undeveloped sense-perception. I happened to meet this person whose index I knew to be crooked after he had gone through a psycho-therapeutic treatment of three months and accidentally discovered that it had become perfectly straight. Much effort during the analysis had been devoted to the development of his powers of sense perceptions.

We have learnt in the preceding chapter that thumb and index are bound up with the conscious, and ring finger and little finger with the unconscious spheres. Thus the position of the fingers helps us to find the point of gravity and the main problem of the personality. Thumb and index reveal the conscious attitude to the world which is connected with the will, whereas the third and fourth fingers disclose the intellectual-spiritual aspirations of a person. The middle finger inclines to the side which harbours the centre of gravity of the personality. It is a fact that thumb and index, which are allotted the task of penetrating the surrounding world, are less revealing as regards talents than the ring finger and the little finger which are intimately connected with and related to the personality and influenced from within.

Musicians will confirm that thumb and index are naturally nimbler and much better suited to technical exercises on the piano than the third and fourth fingers which need considerable training and much practice before they obey the will of the music-maker.

If there is a marked difference in the position of index and middle finger (see fig. 47) as compared to that of ring finger and little finger, that is to say if the first and second fingers have a more outward and the third and fourth a definitely inward bend, there is a discrepancy in the external and internal attitudes of the person as well as a divergency in his conscious and unconscious attitudes. Besides, it is possible that this position indicates a discrepancy between the professional and the personal life.

Mostly such persons are certain, self-assured and positive in their professional domain but wavering, uncertain and afraid to make decisions in their personal life. Combined with certain signs of the inner hand this position can also indicate a disposition to melancholia.

Also the three phalanges of the fingers which are best studied at their inner side are of vital importance. They are clearly articulated by two grooved lines at the beginning and end of the middle phalange (see fig. 63). The lower phalange, which is the one directly linked with the back of the hand, represents materiality in an unintellectual hand and rationality in a more intellectual hand. The second or middle phalange indicates spirituality, by which I mean a combination of intellect and imagination. The upper phalange, which is accentuated by the nail, stands for imagination and feeling. If one of the phalanges is particularly long the quality indicated by this phalange accordingly preponderates within the personality. Equal length of the three phalanges discloses a good equilibrium of the qualities they represent. It must be recalled that with the fingers, and their subdivisions, only the functions of the psyche are connected and not its reactions.

If the thumb is rigid and its upper phalange has an outward bend and the thumb is closely leaning towards the hand, while the index is turning away from the other fingers which have an inward bend (see fig. 48), we infer that the person has a strong desire for self-assertion without adequate inner independence, and that he is egocentric, narcissist, and socially unadapted. There is a great discrepancy between pretension and the power of realising this pretension which may constitute a pathological danger.

A thumb which stands out well—indicating self-reliance—and has a good phalange of will-power but is relatively too small, suggests a person who overstrains his powers, is always in motion and cannot wait and bide his time because his vitality and strength are not adequate.

An index which is not completely conical but slightly rectangular and quite straight indicates bad adaptability (compare Chapter Four), and among other things a talent for organisation. The fact that good organisers as a rule are badly adapted has its explanation in the fact that their powers of observation are applied to a limited domain, whereas a generally good observer is so fascinated and preoccupied with his surroundings that he cannot achieve the degree of concentration essential for organising.

If the index is straight and its upper phalange has a slightly outward bend (see fig. 27) the optical sensibility is good. If this bend is strong and the finger stands slightly apart from the other fingers and is very long, curiosity is indicated.

When the index and the ring finger are inclined towards the middle finger (see fig. 49) without being crooked, and the three fingers are arranged at equal distances from each other, we conclude that the person has a pronounced sense of justice which is often combined with great sensitiveness to injustice.

A definite bend of the second and third fingers towards the index (see fig. 50) shows that the centre of gravity of the personality lies within the conscious sphere. Matters and things of the surrounding world of practical realities are overestimated and overstressed, the activity of the personality is concentrated on the external, and the inner life is neglected. If the position of the index and the middle finger tends to the third and fourth fingers the demands and necessities of the world of realities are treated with neglect and everything that is connected with the self is overestimated.

When the index is straight and knotty the person is a good observer, though only in a limited sense, for he observes, subjectively and egocentrically, merely that which is of special interest to him and considered worthy of concentration, whereas other phenomena of the surrounding world of realities which are taken in by the generally good observer are disregarded. Such people try to include and incorporate the observed phenomena into the range of their intellectual perception.

If the index strives away from the other fingers in the direction of the thumb (see fig. 26), it may be that its owner listens to the opinions and views uttered by his fellow beings, pretends to agree with them but does not really do so, which means that his social adaptability is not genuine. Should such a branching off index be also expressive of keen powers of observation, the bad adaptability—which is suggested by the branching off—manifests itself in the fact that the person is very anxious to know the effect he has on the people of his environment and is always suspicious that he does not make a favourable impression, and consequently lives in a state of constant inner tension, being too dependent upon his environment.

If the first, second, third and fourth fingers have a strong inward bend and the index is placed slightly apart from the other fingers (see fig. 51), we know that the person involuntarily and comparatively easily allows himself to be distracted from his innermost aspirations by external forces. The attitude of these people is often produced by a sense of duty as, for instance, in the case of women who have intellectual interests and an urge for inner development but frequently forgo these for the sake of domestic duties.

If the four fingers of the uplifted hand are closely held together (see fig. 52) we infer that the person is strongly inhibited and often also afraid of the opinions of others, that he is relying on conventions and finds it difficult to express himself. When the fingers are also very long and thin the nervous constitution of the person is poor.

When the four fingers are clearly set off from each other (see fig. 43) the personality is free from constraint and is averse to compulsion and conventions. If the relationship of these fingers is good, that is to say, if they stand at equal distances from each other, the relationship of the various functions is a harmonious one which furnishes the possibility of an all-round development of the talents and dispositions of a person. A middle finger, however, which is strongly inclined towards the ring finger (see fig. 53), is an indication that the intellectual processes are strongly influenced by the emotional-intuitive forces. Reversely, if the ring finger is strongly inclined to the middle finger (see fig. 54), the emotional-intuitive are strongly influenced by the intellectual-rational forces. Such people are usually emotionally uncertain and have little confidence in their original, elemental powers. This may produce strong inhibitions in the personal as well as in the professional life, which often develops into serious inferiority complexes. In many cases such inhibitions can, of course, be bettered or removed by the right psychotherapeutic treatment, and the discovery and explanation of their original causes. Anyhow, the strong inclination of the ring finger towards the middle finger is an unmistakable symptom of a lack of confidence in the emotional powers.

An upper phalange of the middle finger with a slight outward bend suggests a person who has a certain mediumistic quality and who has presentiments about experiences and future events. These states which produce presentiments and premonitions are brought about quite unconsciously and do not demand a special effort of concentration. If, however, the third finger has the same outward bend (see fig. 45), the person possesses what I should call mediumistic knowledge which manifests itself when creative forces are required, and is produced by very intensive conscious concentration.

We have already discussed the jutting-out little finger in the preceding chapter, but there are a few traits which have to be elaborated or added. It is very often combined with a jutting-out or set-off index which, as already mentioned, indicates bad adaptability and is often coupled with a difficult partner-relationship. The social isolation which is an effect of bad adaptability to reality is unconsciously overcompensated for by exaggerated demands on the partner, or, in other words, these isolated people project or concentrate all the libido which is normally claimed for social adaptation upon their partner. The resulting strain is bound to lead in the course of time to serious difficulties. The lack of adaptability expressed by the set-off index can also produce other negative results in the partner-relationship, for the isolation fosters a feeling of uncertainty which becomes a constant source of distrust, jealousy, and a wavering between attraction and repulsion.

If the little finger has a very knotty lower phalange, a conical upper phalange (see fig. 31) and expresses—like the rest of the hand—sensibility we can conclude that the person is erotically very responsive and receptive, but impeded and inhibited by a critical attitude where the realisation of the relationship is concerned. The objective, theoretical and generally human interest in all matters of the partner-relationship is very pronounced and there is a strong, sometimes unconscious, tendency to place the friendship and the intellectual-spiritual part of the relationship higher than the erotic. Such a person has the capacity objectively to look at his own partner-relationship, an inherent quality which induces other people who wish to unburden their hearts to confide their difficulties to him. The above-mentioned inhibition appears only when this type of little finger is also deeply rooted. If this deeply set-in little finger (compare the preceding chapter) has a very knotty and thickened lower section and at the same time appears somewhat rigid in expression (see fig. 55), this is usually the indication of a strong mother fixation which results in inhibitions, fear and uncertainty in the sexual sphere combined with a fear of shouldering responsibility. Such people either find their partner late in life or they indulge in superficial, ever-changing affairs which entail no human responsibility or obligation. Inhibition or fear in the partner-relationship sometimes produces symptoms of psychical impotence. It need hardly be pointed out that all these inferences based on the little finger must be supported by the expression of the hand as a whole and the analysis of the Mount of Venus. A little finger with an inward bend (see fig. 25) suggests a certain inscrutability, and secretiveness and sometimes an ambiguity in the partner-relationship.

If the upper phalange of the little finger has an outward bend (see fig. 45), the person is extremely sensitive to noise and hypersensitive to disagreeable noises. This type of person, who can be disturbed or upset by the slightest noise, has mostly a very keen sense of hearing, and is usually definitely intellectual. He is more at the mercy of influences and unexpected reactions of his environment than the intuitive type who can shut himself off from the world of external realities and thus protect himself better than the brain worker.

If the little finger is in accord with the other fingers of the uplifted hand but clearly stands apart after the relaxed dropping of the hands, our conclusion is that the person is trying to hide the fact that there is a problem in this direction by making his neighbours believe that this particular part of his personality is in order.

The preceding exposition has shown us that the relative position and comparison of the fingers as well as the position and shape of each individual finger permit of fairly far-reaching conclusions regarding the dynamic possibilities for the development of the personality. We must, however, recall what has been said in the third chapter about the differentiation between outer and inner hand, according to which the outer hand with the fingers reveals the static and the inner hand the dynamic properties. Thus we must always keep in mind the chirological principle that the described phenomena are tendencies and dispositions which can either be transformed and changed or remain unchanged. The budding chirologist must be extremely careful in the estimation and evaluation of these particular phenomena and avoid making pronouncements before having examined the inner hand as well. On the other hand, the experienced and psychologically well-trained chiro-analyst can glean from the position of the fingers and their relationship to each other valuable clues for ascertaining the most essential points of the central problem of the person seeking advice.

Nails and Moons

We have already learned in our discussion of the various fingers that the nails provide valuable clues for diagnosing pathological dispositions or weaknesses of certain organs, but I must warn the student of chirology to proceed with the greatest caution when ascertaining his facts and when conveying them to the person concerned. Never should a single sign be deemed sufficient to draw a definite conclusion as to the nature of an illness, and even if several signs point in the same direction, which justifies the assumption that a certain organ is definitely weak, the person seeking advice must be told in a manner which does not alarm him unnecessarily, for as we have pointed out before, persons who consult a chirologist are most highly suggestible, especially as regards organic diseases. It cannot be repeated too often that it is not the task of the chirologist to diagnose illnesses but to point out weaknesses of disposition and thus to make it possible for prophylactic steps to be taken in time. It is an entirely different matter if a doctor wishes to have the opinion of a psycho-chirologist in a particularly puzzling case. The doctor, of course, can be told without reservation what might be dangerous to tell to a person who is not in the care of a physician, and I am glad to say that much valuable collaboration has already been achieved in this respect. The form and colour of the nails are not only a means of diagnosing illnesses of the body, they also have a psychical significance.

We distinguish between long nails which are curved lengthwise and breadthwise (see fig. 56), and long nails which are more broad and not curved (see fig. 57), nails of medium length which are more broad than narrow (see fig. 58), and medium nails which are narrowing down (see fig. 59), and finally short or small nails (see figs. 60 and 61). The length of the nails is not absolute but must be judged according to the length of the fingers.

The long, curved nail shows a disposition to diseases of the lungs, the long and more broad nail indicates a disposition to illnesses of the bronchia and bronchial catarrh. The nail of medium length when more broad than narrow shows a disposition to illnesses of the pharynx and the tonsils. Short nails reveal a disposition to cardiac diseases, particularly if the nails have no moons. If the nails are short and the little finger is definitely crooked the person may have a disposition to abdominal diseases.

These various shapes of nails, however, do not indicate that the people who have them are going to contract these illnesses, but only that the organs concerned are weak. Often these illnesses appear in childhood and disappear later on when the whole organism is fully developed. Moreover, the dispositions are only effective if the nails of *all* fingers have the same form. Whether there is an actual danger of falling ill can only be gathered from the consistency of the nails, the colour being one of the most important features in this respect.

The normal colour of the nails which is symptomatic of good health is a light pink. If all the nails are pale or white, the person is generally anæmic. The pale colour of one particular nail suggests that the organ is affected which is represented by that particular finger (compare the preceding chapter). If the nails are strongly " ribbed ", lengthwise as well as breadthwise, and moreover of a pale colour, we can conclude with a fair measure of certainty that an illness is in the process of starting.

When white dots appear on the nails we must find out whether they have always been there or become visible only occasionally. The latter case denotes a state of great strain and fatigue. If such people rest and recover the dots disappear. Should the white dots be permanent, this indicates a local derangement of the alimentary functions usually in connection with inner secretory disturbances. They are often a premonitory sign of rheumatism.

Red nails reveal that the composition of the blood is not adequate, that it is lacking in phosphorus, calcium or iron, or that the distribution of red and white corpuscles is bad. Sometimes they are symptomatic of too high blood pressure.

Bluish nails point at vascular trouble, circulatory disturbances and a weak heart. By vascular trouble we mean irregularities of the circulation of the blood. The ducts and canals do not work evenly, with the result that because of a spasm produced by cold, one part of the hand appears red and the other pale. This disposition to vascular trouble can also be recognised by the fact that the middle of the top phalanges just before the beginning of the nails appears indented (see fig. 62).

The psychical function of the nails is again interconnected with the physical. Long nails, for instance, suggest emotional, optimistic people who are not very critical, who live more in the realm of imagination and are receptive of irrational impressions. This disposition is characteristic of many sufferers from lung diseases who even in the last phase of their illness make phantastic plans for long periods ahead. By indulging in these phantasies they protect themselves against the sad realism of their illness and the hopelessness of their position.

Short nails indicate sceptical, critical, suspicious people who have a great desire for rational thought and discussion. Here again the psychological disposition acts as a safeguard for the physical side, since it protects the sceptical, rationally minded person from a multitude of impressions and thus reduces the expenditure of physical energy. Short and somewhat broad nails are characteristic of a self-willed, quarrelsome person.

The semi-circular shapes extending from the base of the nails, called moons, indicate what kind of blood circulation a person has.

The thumb nail which indicates the consistency of the blood has always a moon. If the moons are missing on the four fingers or if they are too small the circulation of the blood is poor or irregular, or the blood pressure is too low. When the moons are too large the vascular nerves are hyper-sensitive, the circulation of the blood is irregular, the person is altogether too excitable and impetuous. Combined with certain lines of the inner hands, which will be discussed in due course, too large moons sometimes indicate an irregularly functioning thyroid gland. Irregular moons—that is to say if, for example, two fingers have too large moons the third a small one and the fourth none—are usually an indication of some glandular trouble. They also suggest hyper-sensitiveness as regards climatic conditions and extreme temperatures as well as a great changeability of moods.

CHAPTER SIX

THE SIGNIFICANCE OF THE RIGHT AND LEFT HANDS

As already stated in the second chapter of this book, my method of psycho-chirological interpretation differs from all other chirological methods chiefly on account of my inclusion of the influence of ancestors and parents as an essential part of the synthesis of a personality. This particular point is based on the interpretative significance of the right and left hands, for I consider and treat the right hand as the ancestral hand and the left hand as the hand of individual personality. All the methods known to me claim to see in the left hand the dispositions and talents of a person and in the right hand what he has made of them, they see experiences and developments of a personality up to the thirtieth year in the left and from thirty onwards in the right palm of the hand. Though these two assertions are inaccurate and psychologically unfounded, it is a fact that hand-readers who base their work on these two assumptions nevertheless obtain correct results, at least as far as the static features of the hand are concerned. How then is it possible to reconcile this discrepancy between incorrect methods and partly correct results ?

As the traditional chiromantic and chirological methods are mainly concerned with the ascertaining of static (not psychological or analytical) qualities of the character and confine their pronouncements to events and predictions of events, it is obvious that the observations of such hand-readers are drawn from types of persons who are only interested in static facts and events and, therefore, must be of a more or less undifferentiated intellectual and spiritual status and development. I have found by experience that the right and left palms of people who are less capable of intellectual and spiritual development mostly agree in their arrangement of lines, a phenomenon that will be dealt with later on. Thus the right hand as well as the left hand of such people will yield similar results, provided one does not, as I happen to do, attribute greater importance to the rôle played by the preceding generations. My observations, however, are chiefly based on people who have had to struggle with more or less serious inner difficulties, that is to say, people who are more or less capable of inner development and, therefore, mostly subject to inner conflicts and crises. The right and left hands of such people are different and, therefore, must be interpreted differently.

I hit upon my new interpretation of the hands quite independently and by way of practical experience when I was a very young man. I had become acquainted with chirology through the lecture of a doctor, and according to his teaching I interpreted the two hands in the traditional manner but tried to check up my findings by asking the people whose hands I studied to what extent my observations were correct. They confirmed that my findings were true though many of the things I had told them did not concern themselves but either their parents or their grandparents. For years this ever-recurring reservation kept puzzling my mind till one day I began to inquire which hand it was that provided me with facts relating to the parents or grandparents only. It did not take me long to ascertain that it was invariably the right hand, and in the course of years I realised the insufficiency of the old interpretation, and having accumulated sufficient material I established my own theory. My new interpretation was confirmed by the examination of a number of wounded soldiers who had lost their left hands, for their right hands never revealed personal data or individual qualities, but everything that pointed to their ancestors. A further confirmation of my theory will be found in the fact that whereas the lines and signs of the right hand very rarely change in a lifetime, those of the left hand do so fairly often.

The right hand is the ancestral hand in a biological as well as in a psychological sense, for it shows qualities, reactions and experiences of those ancestors with whom the person is biologically and psychologically connected. The connection is biologically effective if all or part of the lines of the right palm agree with those of the left. If this is the case, the traits revealed by the lines are hereditary. If the lines of the inner hands are different, the connection is of a psychological nature. The right hand of a person does not, however, show a complete picture of the parents or their objective portrait but its subjective counterpart as seen through the eyes of the person, what Jung describes as *imago*.

A complete agreement of shape and course of the lines in both hands also means that there is little scope for inner development of the personality. This state could be described as a deadlock of a generation, as unborn individuality. This does not imply that such people are not efficient in their work ; on the contrary, they may achieve much. They are, however, incapable of intellectual and spiritual development and remain, in this respect, primitive and infantile. This situation is often found in the hands of people whose ancestors have spent their energies and vitality without restraint, and in order to restore the balance the development of the person is suspended for one generation.

A matter of an entirely different nature is the absence of a vitally important line or some other anomaly in the right hand which has its exact counterpart in the left hand. Whereas the agreement of lines in both hands is a normal occurrence, this phenomenon is the sign of an inherited, and therefore in most cases, incurable anomaly. Moreover, it provides an invaluable and unique means for the experienced chiro-analyst to ascertain whether an abnormal trait is constitutional or psychologically conditioned. I had ample opportunity to test and confirm this discovery by studying the hands of patients of various lunatic asylums. If the traits of abnormality—usually the absence of a vital line—appeared in both hands the diagnosis and prognosis was " incurable". I found, however, that in spite of a complete agreement of the abnormal traits in both hands the actual illnesses or diseases were not necessarily the same. Thus, for instance, the son of a habitual drunkard was suffering from an organic brain disease, and a serious psycho-pathic case was the son of a father who had committed acts of violence. I would stress once more that in order to ascertain whether an anomaly is constitutional and incurable or psychological and, therefore, possibly curable, it is essential to find out whether the lines indicating the anomaly are missing in both hands or only in one.

A few examples will illustrate the case of a merely psychological, and not biological, connection between the two hands. A line, for instance, in the right hand shows that one parent was suffering from depression ; the left hand, however, reveals that the person does not share this disposition to depression and yet he is actually subject to moods of dejection. This apparently contradictory position can be easily explained : the person, though not constitutionally disposed to depression, is psychologically and mentally so strongly influenced by the parent concerned that he is now convinced that he is constitutionally suffering from the same complaint. The experienced chirologist, however, can reassure him that this is not the case but that he merely psychologically identifies himself with the parent because of a milieu-fixation. The elimination of this wrong conviction, beneficial in itself, often leads to an improvement in the state of mind of the person.

A further psychological connection between the right and left hands is constituted by the fact that the right hand shows the problems of the parents which they were unable to solve. These unsolved problems would influence the person and play their, at first unconscious, part in the dynamic make-up of the personality. Of course only those experiences and eventful periods of the parents or ancestors which affected him and produced a response are reflected in the right hand of a person. This is proved by the fact that the right hands of brothers and sisters usually record different experiences and events in the life of their parents. There are cases when the right hand of a woman shows that her mother had suppressed part of her own libido and, therefore, had never been able to fashion her life according to her own desire. Then she transferred her libido to her child. It often happens that the partner-experiences of the daughter of such a woman—provided the lines of the left hand are quite different—always come to a negative conclusion. These affairs are often accompanied by a strong feeling of guilt which leads to a defensive attitude towards the other sex, but in spite, or perhaps because, of these feelings of guilt, new relationships are entered into with equally negative results. This goes on until the daughter has "lived off" the repressed and unlived experiences of her mother, and only after she has gone through this "substitute life" is she free to live her own positive experiences which are in keeping with her own, individual self.

In other cases the right hand shows great difficulties in the married life of the parents or a divorce. The left hand, however, reveals signs of greater harmony but also a fear of life, or a fear of relationship with the other sex, escape from reality or flight into unsatisfying work or, as a counteraction, into a premature marriage. Here too a comparison of the two differently lined hands made it possible to prove that all these reactions were not constitutionally conditioned but could be traced back to influence from outside, and an unconscious, negative connection with the milieu.

As already mentioned, the right hand contains the picture of the parents as seen by the owner of the hand, a portrait which is bound to be subjective and incomplete. This subjective picture and the corresponding attitude of the person concerned produce and constitute the parent problem. The resulting fixation plays a decisive part in the development of the person and strongly affects his or her attitude to and relationship with the opposite sex. As I shall refer to this attachment as a milieu-fixation I should like to give a short definition of this term.

By milieu-fixation I mean that a part of the forces and energies of the mind has been arrested and is being prevented from freely unfolding itself within the personality by an undeveloped, infantile, and mostly unconscious attitude to the parents. The signs and lines of the inner hand tell us of what nature this fixation is.

Before going further I must say something about the procedure I adopt when analysing hands. After the analysis of the outside of the hands I always begin with the right inner hand. Its marks and lines supply me with the qualities, the experiences of the parents as well as the periods when they happened, and I ask the person whether my findings concern the father or the mother, or both. If the negative signs predominate in the right hand and I am told that they apply to the mother, my conclusion is that the person has a negative mother fixation, that is to say an opposition to the mother. In case the father is named, we are dealing with a negative father fixation. Should the positive traits preponderate and refer to the mother whereas the negative ones are attributed to the father, we have a positive mother fixation and a negative father fixation which will induce the person to overestimate the mother and underrate, and therefore disdain, the father. Both positive and negative traits in the right hand attributed to father and mother, show a fairly objective attitude towards the parents, greater independence and more scope for individual and personal development. Apart from the already mentioned fact that brothers and sisters always differ in their right hand because each child sees the parents in a different light, we must also take into consideration that parents for reasons of which they are unconscious behave differently to their various children, even if they have the best intention of treating them with equal justice. Moreover, the children were born and are growing up at different periods in the development of their parents.

I remember a case of two sisters which is a striking example of the completely subjective attitude towards the same parents. Their right hands contained signs that one of the parents must have been a domineering and cold-hearted person. The one sister was positive that it was the mother, whereas the younger girl was equally emphatic that it was the father, and no argument would induce them to correct their contradictory conviction. Although the attitude of children to their parents is subjectively coloured, it nevertheless leads to definite difficulties and later on in life produces problems of which we can give only a few examples.

The more striking the difference between the right and left hands, the greater the difference of the person's attitude, character qualities and talents as far as heredity is concerned. If, in spite of great differences, certain similarities also appear, it often happens that the person displays a negative and hostile attitude to these traits simply because he shares them with his parents. Through this opposition, however, an unimportant matter becomes a main issue, and the person is impelled to concentrate his energies to a considerable extent on something negative, thus neglecting his own individuality which is composed of the qualities not prevalent in his parents. Quite frequently a comparison of the right and left hands discloses that sons and daughters in their opposition to or even hatred of a parent, practise the very thing they profess to loathe in their parents, in the erroneous conviction that these reactions are constitutional, and unconscious of the fact that they are actually conditioned by an antagonistic impulse or hatred. If, for instance, one of the parents happens to be a domineering and despotic type and the owner of the hand has learnt to despise this attitude without, however, being quite free of it himself, it may well be that the suppression of this character disposition combined with the opposition to the parent concerned may lead to neurotic complications. Such complications will develop more easily if the child is predisposed to them by instability and a high degree of differentiation. It is also most important to take into consideration to what extent he is conscious or unconscious of the cause of his difficulties. The chiro-analyst, however, by a comparison of the two hands, is in a position to find the central problem of the person in regard to his milieu-fixation and his attitude towards his parents, thus opening the road to therapeutic treatment. Then the psychotherapeutist by reconstructing a fairly objective and comprehensive portrait of the parent can proceed to release the analysed person from his fixation. Only an effective liberation from such a fixation provides a possibility of living on one's own physical and psychical resources.

Undoubtedly this is one of the central problems of our generation which can only be solved by the creation of a strong inner foundation and inner balance. This problem looms particularly large in our time because the difference in the conception of life among the contemporary generations is greater and the dynamism and the tempo of life are more divergent than ever. Moreover as a result of the " re-valuation of values ", the religious basis which sustained previous generations, and the belief in capital as a stabilised value, have nowadays more or less vanished. The only palpable support is life itself relying on its own vital energies. To fulfil this demand, it is more than ever necessary that the forces and energies of life should be free from all impediments and brakes such as milieu-fixations, wrong education and the affective attitude resulting from them.

A few instances of the negative effect of strong parent-fixations, which I come across almost daily in my chirological practice, will suffice to illustrate my point. A man with a strong mother fixation is always trying to find in every woman his mother, his ideal of womanhood, a quest which inevitably results in disappointment because he is not prepared to give up his mother or rather his unconscious mother fixation. Later in life, according to his general disposition, he either has many experiences with women or marries prematurely. In the first case he is afraid of a lasting relationship and sabotages it by making exaggerated demands on all the women he meets, not knowing that his decisions are guided by his unconscious mother fixation which renders him incapable of a harmonious partner-relationship. In the second case, the man unconsciously sees in his wife a substitute for his mother and accordingly wishes her to play this part. This position is more serious in its consequences for the wife, because this enforced and unnatural part must leave her dissatisfied, which leads to complications and difficulties in their partner-relationship.

A man who is opposed to his mother because of his negative

mother fixation unconsciously transfers his opposition to all other women and according to his general disposition tends to become homosexual.

The result of a one-sided positive fixation of a daughter to her father are unreasonably high demands on the personality of a possible partner, demands which no man could be expected to fulfil. Unconsciously such a girl would always contrive to meet men who are sure to fall short of her expectations, for the father to whom she is so strongly attached must not be dethroned. The difficulty here is not so much the establishing of a relationship but its development and fulfilment, because the idealised father or father *imago* by which essential parts of her libido are bound unconsciously intervenes. These exaggerated demands on the partner usually lead to a negative conclusion of the partner-relationship and also to a multitude of experiences. If, however, the father fixation is strongly negative, her critical attitude to her father becomes the cause of a negation of everything that is connected with the opposite sex. By the negation of her father she is bound to attract men with similarly negative traits who never fail to disappoint her and who thus provide her with reasons to despise them as she despises her father. Her life is a continuous search for the ideal father who was denied her and after a chain of disappointing experiences she turns away from man altogether to her own sex to whom her critical attitude does not apply.

Finally the comparison of the right with the left hand shows the true inner task or mission of a person, for apart from the plain duty of doing one's job in accordance with one's particular talent and suitability, every human being who is capable of development has also to carry out an individual task which cannot be accomplished by anybody else. This task is closely connected with the lives of the preceding generations, with the failure of his forefathers to solve certain problems, with their avoidance of experiences or their passing through the wrong experiences. This individual task which can be ascertained by a comparison of the two hands makes the life of an individual a necessity and gives it a meaning in the eternal cycle of the generations.

A few cases may illustrate this. In the right hand of a woman of about fifty, I could see that her parents were only concerned with and interested in the material, practical, soberly rational side of life. The lady confirmed this when she told me that her father's ancestors had been bankers and merchants for centuries. Her left hand, however, clearly disclosed her divergence from the family tradition, revealing that her interests were entirely intellectual and spiritual with a marked philosophical-religious trend. In the course of her development she had to overcome countless difficulties and obstacles and had passed through periods of utter dejection, of uncertainty accompanied by inferiority feelings and lack of faith in her personal mission and destiny. After a comparison of her hands I was able to explain to her that she encountered these unusually great difficulties because it was her task to change the direction of the tradition of her family and within one generation to effect the switching over from a materialist tendency to a more inward and spiritual attitude to life. Now that she had become conscious of the nature and realised the inner meaning of her life she was able to continue her struggle with fresh enthusiasm, to endure and to overcome her difficulties with greater ease.

I recall in this connection the case of a woman of forty, an artist whose right hand revealed that one of her parents, though endowed with great artistic gifts, was extremely vain and exclusively interested in external success and the applause of the public. This negative trait in the character of the parent—it was her father—aroused the strongest disapproval of the daughter. Her problem and difficulty consisted in the conflict between her artistic aspirations which were concerned with her inner development and the attainment of perfection, and a desire for recognition and external success. It was plainly her task by consciously concentrating on her inner spiritual development to purify and to perfect the artistic and creative element which in her father's development had been thwarted by his extreme extravert attitude. It was because of her wavering between these two tendencies that she jeopardised her chances of external success. Once she had realised that it was the task of her generation to attain the inner development, recognition from outside was almost automatically achieved.

The right hand of a girl of twenty-two, the descendant of an old aristocratic family, showed that her ancestors had lived a life of convention and rigid tradition which had led to a standstill in the development. The girl was a sculptor and thus had effected a break in the tradition of her aristocratic ancestors.

Apart from the difficulty of overcoming the opposition of her parents to her artistic career she suffered from strong inhibitions which assumed the expression of ever-recurring doubts of her undeniable talent. Periods of deep depression, discouragement which led to extensive interruptions of her artistic work were the result. Also in her case the realisation of the fact that her great difficulties were not merely an individual and personal matter but caused by the tremendous task of breaking with a family tradition and of initiating an entirely new line of activity greatly facilitated the successful performance of her task.

In this context I should like to mention that children whose parents have lived an outward and completely unconscious life are allotted the task of living an inward and more conscious life which is often responsible for difficulties in the development of their personalities. If such a person fails to complete the task set for him by the preceding generation the obligation passes on to the succeeding generation.

Quite frequently one finds the same signs of an experience in the right and in the left hand as well as an indication that the event took place at the same age. Thus, for instance, the right hand shows that the parents were divorced at the age of forty and the left hand reveals that the marriage of the owner of the hand was dissolved also when he was about forty. This usually happens if the problems concerned were not properly assimilated by the preceding generation, that is to say the parents had failed to get over their decisive and negative experiences, had failed to adjust them to their general development and were caught in them, being incapable of grasping their meaning or recognising their necessity. Consequently the owner of the hand must once more go through these experiences in order to understand and to realise them.

Thus we see in the right hand an integral and inseparable part of the personality as well as a means of finding the key to the difficulties of a person and their cure, always provided they are not constitutionally conditioned. This aspect of the interpretation of the hand, that is to say the correct position and application of the features of the right hand within the range of the individuality of the person, is in a way the most essential part of chirological interpretation, because it leads from analysis to synthesis, to therapeutic treatment and cure.

In this respect the work of the experienced and responsible

chiro-analyst can be of great social importance, since his insight into the connection between the owner of the hand and his ancestors enables him to reveal to the analysed person the innermost task of his life, thus helping him to find his appropriate place in the social community, that reservoir of dynamic forces from which all civilisation and progress spring.

CHAPTER SEVEN

THE MOUNTS

The Mounts are elevations or swellings below the fingers revealing how the functions which can be observed in connection with the fingers are translated into active application within the mind. They are as it were reservoirs of power confirming qualities and talents which we have seen expressed in the fingers and shall meet again in connection with the lines of the hand. The essential point about the Mounts is that they express action.

An interesting historical confirmation of the active character of the Mounts is the fact that in the chirological writings in the literature of India, dating from the Vedic period, that is to say about 2000 B.C., the Mounts as a feature of the human hand are unknown. This agrees with the generally accepted idea that Indian civilisation and the Indian attitude to life have an inward tendency with a strong trend to contemplation, perhaps best known through the doctrine and practice of Yoga, where all activity is transferred from the outward to the inward. The Mounts are mentioned for the first time in the history of chirology in the works of the ancient Greek civilisation, which did not believe in the training of the mind and spirit in isolation but was also interested in the external phenomena of life, in sport and even warfare. Greek mentality and culture were obviously more concerned with the activities of the outside world than the Indian. Even to-day the hands of many pure Indians do not show the Mounts which are generally found in those of Europeans. Their absence in the hands of non-Indians is always a sign of passiveness.

A very essential point in the interpretation of the hand as a whole is the size and the consistency of the Mounts. Mounts of normal size, that is to say, in adequate proportion to the hand, indicate a positive quality; too large Mounts have a negative significance, and the lack of Mounts, as we have already said, the absence of the active manifestation of the quality concerned. Furthermore we must consider the relative size of the individual Mounts whether they are large, small or too large. The absence of a Mount, however, does not mean that the quality is lacking altogether; it merely suggests that the quality is not translated into action. An exception in this respect is the Mount below the thumb and the opposite Mount at the lower outer edge of the hand.

These two Mounts, which will be discussed later on, are always in the same position, whereas that of the others below the fingers is subject to greater or slighter alterations. This shifting of the position of the Mounts which can be observed in every hand is explained by the fact that human activity is not rigidly and unswervingly set in one direction but is manifold and finds its outlet in various ways which influence each other. Once more I should like to point out in this context that every feature should be interpreted in connection with all the other features of the hand as, for instance, the general expression of the hand, the fingers, lines and Mounts, each representing a different aspect of the psyche. Thus an isolated interpretation of the Mounts would yield only a distorted or fragmentary portrait of the character.

In conversations and discussions with scientists I have had many opportunities to observe that they strongly object to the traditional, mythological denomination of the Mounts introduced by the ancient Greeks and accepted by all chirological textbooks, regarding these terms as sufficient proof of the mystical and therefore unscientific character of chirology. When we remember that mankind always seems to have an immanent desire for establishing contact with divine forces and the universe, and that the significance of the Mounts agreed with the ancient people's conception of those particular gods, the argument of the scientists seems unduly exaggerated and I see no reason why I should not use the already established terminology.

The Mount below the base of the thumb forming the ball of the thumb is called the Mount of Venus (see fig. 63, No. 1) and shows the sensual, creative force, the primordial, earthy power which manifests itself in various ways within the psyche. It is the Mount whose features are most strongly connected with the environment. Among all the Mounts the Mount of Venus, which always occupies the same position, is of outstanding importance. It is divided into two sections. The upper one which, however, does not appear in all hands, forms a new elevation between the root of the thumb and the first line across the hand.

The lower part of the Mount, provided it is of normal size,

shows an amiable, warm-hearted person who enjoys the pleasures of life. If it is complemented by a certain Mount which will be discussed later on it indicates an artistically gifted, or artistically interested, person. A weak Mount or its entire absence suggests the absence of these qualities, and we may assume that such a person is either lacking in vivacity and without passion, or if this is supported by other features, is also emotionally cold.

A very strong and big Mount of Venus with no lines on it tells us that the energies and forces which it represents are fully concentrated on the sexual sphere, undifferentiated and according to the size and shape of the Mount, overstressed ; persons with such a Mount are asocial in these matters, sometimes to a degree of callousness. The absence of the lines on the lower part of the Mount signifies, as already pointed out, an undifferentiated accumulation of the above-mentioned forces. The lines which should traverse the Mount can be either horizontal or vertically ascending (see fig. 64). Horizontal lines suggest greater differentiation but the energies of the Mount are put to an egoistic use or are not used at all, whereas the ascending lines reveal the possibility of an application of these forces to other spheres of the psyche which are not exclusively concerned with the body but with art, and a generous colourful attitude to, and mode of, life.

The upper section of the Mount of Venus (see fig. 63, No. 2) represents the sublimated forces of the Mount, that is to say, forces which are less preoccupied with the ego, resulting in readiness to help others, love of humanity, of children and animals, a gift for nursing, motherliness, and also pedagogical and medical talent according to the type of the hand. If no lines appear on the lower section of the Mount and many on the upper we deduce that the person is repressing his instincts and impulses. Thus we have to distinguish between the positive significance of the Mount in the case of genuine sublimation and the negative effect in the case of repression. We speak of repression when the instinctive forces and impulses are partly or entirely and unconsciously transferred to people of the closer and more distant environment, or are completely bottled up instead of being actively and consciously employed. Let us assume that a mother transfers her desires and instinctive urges, which she should release in her relationship with her husband, to her child by surrounding it with exaggerated solicitude and overwhelming it with tenderness and caresses. There would be no sublimation or restraint in such a manifestation of maternal love, but the egoistic seeking of satisfaction of a craving which often is rejected by the normal child.

If the usual lines, horizontal as well as vertical, appear on the lower part of the Mount and on the upper part but not in between, thus producing an area without lines between the two sections (see fig. 64), we conclude that the relationship to the partner is subject to strong, periodical changes brought about by an irresistible urge to be alone and then again to resume the former relationship.

The Mount at the base of the first finger or index, traditionally called the Mount of Jupiter (see fig. 63, No. 3), represents the qualities which are essential to translate aims connected with the outside world into deeds. These I will enumerate : ambition, self-assertion in the personal and in the practical spheres, talent for organisation if this is confirmed by the shape of the finger, family feeling if the type of the hand is pointing in the same direction, in short everything that helps to preserve the human race. Should the Mount be too large and the adjoining middle finger be without a Mount we conclude that ambition, self-assertion and vanity are the dominating features of the personality. People with that kind of Mount of Jupiter are exclusively concerned with their own advantage and know how to make the best use of it. If the Mount is prominent in a fleshy type of hand the person has a strong desire to live a life of comfort, luxury and enjoyment. The absence of the Mount shows a lack of self-assertion and self-confidence which makes it difficult for such persons to carry out their plans, and usually their contact with the outside world is not too good. Many lines across the Mount are a sign of pedagogical talent which, however, is not free from personal ambition and the desire to dominate.

The adjoining Mount situated below the middle finger or slightly shifted towards the third finger is called the Mount of Saturn (see fig. 63, No. 4) and reveals a capacity for philosophy and religious speculation, particularly if the Mount of the Moon which is opposite the Mount of Venus is well developed. Self-tormenting, highly sensitive and misanthropic persons have too large a Mount of Saturn which, combined with a long thin middle finger, often also indicates avarice. The absence of the

Mount of Saturn shows that some power of the psyche is over-emphasised, which prevents a harmonious balance.

The Mount below the third finger which, as a rule, is shifted towards the fourth finger and was known to the Greeks as the Mount of Apollo (see fig. 63, No. 5), is an indication of artistic talent, artistic and aesthetic interests, and, in conjunction with other signs in the hand, of a sense of beauty. Its value is particularly enhanced by a good Mount of Venus. This combination will be found in the hands of nearly all artists who have to rely on and seek the response and approval of the outside world. People whose Mount of Apollo is too strongly developed are predominatingly interested in external things, public success and are easily influenced by flattery.

The Mount of Apollo is frequently combined with the Mount below the fourth finger which suggests that the talents of the person are artistic and scientific, that he is not only artistically gifted but also capable of realising and establishing his talent in the practical sphere. The combination of these two Mounts is almost invariably found in the hands of intuitive scientists and medical men, and often in those of musicians and actors who are dependent on the appreciation of their public.

The Mount located at the base of the little finger which is called the Mount of Mercury (see fig. 63, No. 6), shows, according to the type of the hand, commercial, scientific or political talent. Too strongly developed, this Mount indicates greed, mendacity and a tendency materially to exploit one's fellow men.

The Mount placed on the edge of the palm between the two main lines across the hand, known as the Mount of Mars (see fig. 63, No. 7), is the expression of a person's active readiness to fight. In order to interpret this Mount correctly it is particularly important to consider the general type of the hand. If the hand is of an intellectual or spiritual type, a normally sized Mount of Mars tells us that the person is prepared to fight for an idea, a work, a new method. In a more material type of hand, the Mount shows that the person enjoys sport, fighting in general, the establishing of records. If the hand is of the primitive and completely unintellectual type and possesses a large Mount of Mars we are dealing with a pugnacious, quarrelsome person, a ruffian.

If the Mount of Mars is unusually large and the space up to the three main lines—the Lines of Fate, Head and Heart—is empty, inanimate and devoid of smaller lines (see fig. 65), this is an indication of strong aggressive forces which cannot find an outlet within the personality. As these forces are not being translated into action, the person accumulates feelings of passive resistance and suppressed revolt which very often produce physiological constipation. As its cause is psychological, I am inclined to call it psycho-physical constipation.

The adjoining Mount along the edge of the hand—opposite the Mount of Venus—which more or less extends to the first line across, is the Mount of the Moon (see fig. 63, No. 8). Whereas the Mount of Venus is representative of the physical vitality, its opposite partner, the Mount of the Moon, is solely expressive of the psychic forces of the personality such as imagination, understanding of nature, the unconscious powers, the personal as well as the collective unconscious. According to C. G. Jung the collective unconscious is " the immense heritage of the mind of mankind reborn in each individual brain ". (C. G. Jung, *Secret of the Golden Flower*) or " the wide foundation of a hereditary and general disposition of the human mind which as such is unconscious, the foundation on which our conscious and personal psyche is based " (C. G. Jung, *Die Beziehungen zwischen dem Ich und dem Unbewussten*, Rascher Verlag, Zürich—*The Relationship between the Ego and the Unconscious*). It is the unconscious which is not individually acquired, but has always been present since the beginning of life, being accumulated from the endless chain of our ancestors and shared by each individual ; hence its name " collective unconscious ". (See also Dr. Gerhard Adler's work, *Die Entdeckung der Seele*, Rascher Verlag, Zürich—*The Discovery of the Soul.*) It is chirologically recorded in the lower part of the Mount of the Moon, in the depth of the hand as it were. If ascending lines appear in this part and are joining up with or finding access to one of the main lines of the hand, we can say that the forces and influences of the collective unconscious are applied within the psyche. Should, however, these ascending lines lead into the void without establishing contact with one of the main lines, and should there be a great number of them, then the powers of the collective unconscious find no application and are unable to link up with the more conscious part of the mind. As a result of this the smooth functioning of the nerves is often disturbed by extreme excitability, lack of self-control, changeability, instability or hysteria.

The personal unconscious has a two-fold nature, the first of which consists of experiences we have actually had and have forgotten or repressed. The effect of these, now unconscious, experiences is recorded between the first and second main lines of the hand and will be discussed more fully in the following chapter (see fig. 73). The other aspect of the personal unconscious which, according to C. G. Jung, " cannot be made conscious because so far there exist no bridges of association and relationship for penetration into the conscious sphere " (C. G. Jung, *Über die Energetik der Seele*, Rascher Verlag, Zürich —*On the Energetics of the Soul*) is recorded in the upper part of the Mount extending approximately as far as the first line across. On the lower and more often on the upper part of the Mount of the Moon, we find cross-lines protruding from the edge of the palm into the hand. According to old chiromantic beliefs they are Lines of Travel. It is a fact that people who have many such lines have a definite desire for change which often finds expression in an urge to travel. My enquiries about the goal of these travels and the persons' (mostly ungratified) longing to visit other countries have revealed the interesting fact that in most cases their desire was directed towards the East, which, at bottom, is the unconscious urge to return to the cradle of mankind.

A Mount of the Moon of normal size with only a few lines on it shows a very strong connectedness with nature, and love of animals. A very big Mount of the Moon without any lines whatsoever can mean, according to the type of hand, several things. If the hand is very soft, pale, delicate and has a weak back the person is a drug addict. If the hand is stronger, of reddish colour and has a big Mount of Venus, a large inanimate Mount of the Moon suggests that the person is given to drinking, especially when the little finger is crooked. Such a Mount of the Moon, if combined with certain lines in the hand, can also indicate a person with an overwrought imagination who is possessed by an idea to which he clings with fanatical obstinacy.

Summing up, we can say that a prominent and inanimate Mount of the Moon is expressive of all the desires and instinctive forces which are connected with imagination, phantasy and the unconscious, manifesting themselves in an undirected, one-sided way and without a harmonious connection with the other facets of the personality.

In some hands another Mount immediately above the wrist appears between the Mounts of Venus and of the Moon, the Mount of Neptune (see fig. 63, No. 9). It shows that the person is endowed, as it were, with a sixth sense, with telepathic powers and clairvoyance. Combined with other signs, it is a strong indication of active musical ability.

I wish to repeat once more how important it is to consider the relative sizes of the Mounts, whether they are large, too large, too small or non-existent. A few examples will illustrate the point. If in a soft hand only the Mount of Venus and the Mount below the fourth finger (the Mount of Mercury) appear, this shows a person who possesses many talents, is amiable and perhaps artistically gifted but is inclined to be unstable, inconstant and frivolous. The addition of the Mount of Saturn, below the second finger, for instance, would lend greater stability to the personality and enhance its value.

If a hand which conveys the expression of forcefulness and vitality shows the Mounts below the thumb, the first finger and between the two main lines across (the Mount of Mars), the person has the character of an adventurer who is impelled by his ambitions and sometimes lacks scruples in his methods of achieving his ends.

If the Mount below the second finger is added to this combination the instability and the desire to fight is counteracted by a certain constancy and seriousness.

A hand which has the two Mounts below the thumb and the Mount below the fourth finger and whose general expression is that of materialism, reveals that the person has very strong erotic instincts and a tendency to exploit them commercially. If in addition to this we also find the Mount of Mars (between the two main lines across) the character possesses a note of criminality. As a matter of fact, this combination is very often found in the hands of blackmailers, owners of brothels and calumniators.

In conclusion, I want to stress once more the importance of considering not only the Mounts singly and in relation to each other but also the general character of the hand, its expression and quality.

CHAPTER EIGHT

THE MAIN LINES IN THE PALM OF THE HAND

We have seen that the fingers reveal certain psychical and physiological functions and the mounts the intensity and active manifestation of these functions. The lines of the inner hand, however, reflect the inner reactions of the personality, its dynamism as well as the effect of the environment upon the personality.

Before proceeding with a comprehensive explanation and interpretation of the lines, I should like to emphasise that psycho-chirology neither is nor claims to be an exact or pure science. It is a combination of definitely measurable phenomena—as the size and consistency of the fingers, the back of the hand, the colour of the skin, size, shape and colour of the nails and moons, dimension and course of the lines—and conclusions which, though based upon observation and experience, rely greatly on intuition. I wish to establish this point once and for all, because rationalistically minded scientists who are sometimes inclined to accept the phenomena of the outer hand and their interpretation often refuse to do so where the lines of the inner hand are concerned, maintaining that the interpretation of these lines is achieved by means of telepathy, and that the intuition of the chirologist reads something into them which is not actually there, that the lines are merely the grooves produced by the movements of the hands and consequently phenomena connected with them are outside the range of scientific research. Once more I repeat that I do not claim for chirology the title of an exact science, although I should like to point out that neither medicine nor psychology are entitled to this claim. After all, medicine diagnoses internal illnesses from external physiological conditions and symptoms, and, as in chirology, intuition and speculation come into play. Psychology, for instance, deals with the illnesses of the " soul " and bases its research on the phenomena and manifestations of the human soul, for the existence of which there is no exact proof. Whereas medicine and psychology are universally accepted and permitted to commit errors without their usefulness and necessity being challenged, an infallibility is demanded of the psycho-chirologist which is never expected of the physician, psychiatrist, or psychologist. The experienced chirologist does not neglect the central problem of a case for the sake of its characterological and psychological aspects. If, however, he happens to overlook a detail which the critic deems essential, or worse, makes a mistake, this is often considered as sufficient evidence to condemn chirology altogether. Modern scientific research is so specialised that new facts and data are often unjustly dismissed or accepted because the verdict is based on an insufficient number of cases. After all, scientific research demands the observation and study of innumerable cases before a fact or theory—originally conceived by intuition—is accepted as scientific knowledge. I frankly admit that I cannot offer a definite explanation of the origin of the lines of the hands ; but can the anatomist explain why the heart is on the left and why the other organs of the human body are in their respective places ? In the same way as he looks upon the structure of the body as a given entity, I regard the lines of the hands. According to my own experience I am inclined to maintain that the formation of the lines is directly connected with the central and vegetative nervous systems. I find confirmation of this theory in the fact that after certain inner experiences of an emotional and affective nature, new lines often appear whilst others disappear. Thus I have in my files the prints of the hands of a girl of twenty-one which show in the left palm certain lines indicating a severe conflict with one of her parents, her mother. As the girl was in a critically nervous state I suggested a psycho-therapeutic treatment. A year later I took fresh prints and found that all the lines revealing the conflict with the mother had disappeared. The reason for this astonishing change was, of course, the successful elimination of the conflict during the psychological treatment. This unmistakably shows that there is a close connection between the inner experiences of a person and the formation of lines. We all know how emotional experiences such as joy, excitement, fright, pain and sorrow are reflected in the face, producing changes of colour and expression. As constant sorrow and grief leave their mark in the face in the form of wrinkles and lines, the hand reveals to an even higher degree the imprint of inner experiences because it cannot be influenced and controlled by the conscious mind and is thus capable of greater differentiation of expression.

In this connection it must be pointed out that manual labour and the movements of the hand have not the slightest effect upon its shape, expression and lines, as is often believed by the layman, for only the psyche, the mind, the psychological processes are the formative power. This is clearly shown by the fact that the hands of people doing the same kind of manual labour have quite different shapes and lines. Carpenters, locksmiths, bricklayers, etc., have not more lines than other people or a characteristic shape of hand because they work with their hands. These features solely depend on their psychological and mental make-up. Undifferentiated, undeveloped workers have less expressive hands with fewer lines, whereas workers whose minds are more differentiated and more highly developed have accordingly more expressive and more intricately lined hands. Moreover, the right hand has generally fewer lines than the left in spite of the fact that most of the work is done by the former. The explanation of this may be found in the fact that the right hand is not the hand of the personality but the ancestral hand which reveals the unconscious springs of the conscious part of the personality. Already the palm of the hand of the new-born babe shows the main lines forming, as it were, the basis for the reactions of the personality. These four main lines—for the sake of clarity and to avoid confusion I have retained the traditional denominations—are the Line of Life, the Line of Fate, the Head Line and the Heart Line. With the exception of the Line of Fate these lines must be present in the hand of a normal child. According to my observations the Line of Fate may appear at any time before the twentieth year, and it is not until then that we can decide whether it is definitely missing. We shall hear more about this in the section dealing with the Line of Fate. Recent clinical and hospital research has produced evidence that already in the hands of an embryo of three months —that is at a time when not even the joints of the fingers have been formed—the lines are indicated by thinner stripes in the skin which afterwards develop into proper lines. Anthropologists, contrary to the hitherto prevalent view of scientists that the lines are produced by the movements of the hands and fingers, have come to the conclusion that the lines are a hereditary feature. My own observations of a great number of new-born babies confirm that all children, provided they are not mentally deficient, are born with a greater or smaller number of lines. I have also been able to observe that children who are born with the main lines only and with no—or very few—subsidiary lines are of a more harmonious disposition and have to overcome fewer difficulties in their development than children whose hands also show a great number of subsidiary lines. Subsequently I shall demonstrate in greater detail that these subsidiary lines are the residuum of experiences, impressions and influences of various kinds. If children are born with a great number of subsidiary lines we may assume—and experience has borne it out—that they are burdened with the heritage of the unlived experiences and the unsolved destinies of their parents, a heritage which from the very start constitutes a considerable impediment for the dynamic development of these children.

In analysing the main lines, four important features have to be observed : their course, form, intensity and colour. With the subsidiary lines, which will be discussed in greater detail later on, the starting-point, the end and the relationship with the main lines have to be especially considered. The course of the main lines which is described in the paragraphs dealing with each line is definite, whereas that of the subsidiary lines is not fixed. In fact, they may proceed from any part of the hand, and in order to interpret them correctly it is essential to know exactly where they begin and where they end. To ascertain this we must observe where the subsidiary line in question is tapering off. If a line is thicker at the bottom and gradually tapering off at the top, such a line has an ascending tendency. If, however, the upper part is thicker and the line grows thinner at the bottom, such a line has a descending tendency. The ascending tendency suggests that the subsidiary line has a positive, constructive character capable of development, whereas the descending tendency shows a negative, destructive character impeding development. But let us deal with the main lines first.

Looking at the inner hand, the first line to attract our attention is the *Line of Life*, which according to my interpretation springs from the ball of the thumb, the Mount of Venus, and ends at the first line across, the Head Line (see fig. 66, line *a*). I would like to point out that my interpretation of the course of the Life Line differs from that of all other exponents of chirological systems, according to whom the Life Line starts at the Head Line and terminates in the Mount of Venus. My entirely new interpretation, which is the fruit of many years experience, is mainly based on the fact that the impressions and experiences of childhood, which play an important part in my system, are always to be found in the neighbourhood of the Mount of Venus. Moreover, the Line of Fate, as is universally accepted, commences in the lower part of the hand, and it therefore seems reasonable to suppose that the Life Line should have its starting-point in the same region of the hand for, after all, the fate of a person begins with his entry into life. As a matter of fact I have always found that this disagreement about the starting-point of the Life Line is of little consequence in practical chirology because, with the exception of childhood impressions, all other phenomena connected with the Life Line can be studied equally well at its beginning as at its end. This is a further contribution to my statement made in the preceding chapter that many psychological phenomena can be repeatedly studied at different parts of the hand.

The Line of Life indicates the vitality, the physical power of a person, and in its course are recorded the influences of illnesses and impressions connected with illnesses, accidents or the death of near people. As the Life Line does not reveal the psychical power, the power of the mind, it may well be that in spite of a bad Life Line great possibilities and achievements are feasible provided the thumb and the Line of Fate are good. The Life Line can be described as good if it is not too broad, not too deeply grooved and not too frequently interrupted, but proceeds in a fairly unbroken line, undisturbed, clearly and sharply outlined and compact. A Life Line which is too broadly defined indicates a strong, undifferentiated life urge which is often coupled with flaccidity and inertia. It is often quite erroneously believed that it is possible to ascertain the exact length of a person's life from the Life Line. What can be seen, however, is the intensity of the vitality and the power of physical resistance. Experience has taught me that the calculation of a person's span of life cannot be made with any accuracy, and if at all only by considering the Line of Fate as well. Serious chirologists, however, as I have already pointed out in a previous chapter, should not concern themselves with this question at all, for it belongs to the domain of the fortune-teller. Only in cases of vocational guidance or when considering certain developments or enterprises is it of importance to know something about the physical power of resistance. We shall find this information in the character, course, intensity and outline of the Life Line. But in all cases concerned with physical vitality, we must always take the Line of Fate into consideration, as physical energy and vitality are always bound up with psychological factors.

In ascertaining the nature of a person's vitality and its manifestations, the beginning of the Life Line is important. If the beginning of the line goes round the Mount of Venus (see fig. 67) and thus has its origin in the Mount of Venus, this is an indication that the emotional attitude of a person towards life and consequently a considerable part of his reactions to life, are rooted in the earth. The relationship of such people to nature and earth is usually a very intimate one. Quite frequently they are the descendants of ancient or more recent country families and are more powerfully linked with and more conscious of the life force and its manifestations in them than persons whose Life Line has its starting-point nearer to the Mount of the Moon (see fig. 68). In the latter case the unconscious forces, particularly the collective unconscious forces, which find expression in the Mount of the Moon, have a disturbing and impeding effect upon the life energies. Such people are more desultory and highly strung, unsteadier and more incalculable in the expenditure of their vitality and on the whole more impelled or guided by their instincts and more unstable than the former. This prevalence of the instinctive may manifest itself in any sphere in which the life force asserts itself, also in that of sexuality.

If the initial part of the Life Line is weak and delicate or starts higher up than it normally should, this is an indication of a weak constitution in childhood or a sign that the person has gone through many illnesses when a child. This is a fact of some psychological significance because it very often provides an explanation for a strong inferiority feeling, overcompensated by a craving for power and self-assertion. This inferiority feeling unconsciously springs up in the child which is weaker and sicklier than other children, and thus on account of its delicate constitution feels physically inferior. Later on, when it has become physically fitter and healthier, the acquired inferiority feeling is psychologically balanced by a craving for power and self-assertion. If the Life Line is on the whole much frayed, a certain weakness of the constitution is suggested, mainly caused by the undermining effect of the psychical and unconscious energies on the physical energies. If the Life Line is interrupted, and particularly at a weak spot of the line, we infer a disturbance in the person's health. Such an interruption may, however, indicate merely an important change in the circumstances of life. I have often noticed this phenomenon in the hands of people who had at that particular juncture settled down in a new continent: it is obvious that such a " transplantation " must produce the effect of an incision or break in the life of the person concerned. It indicates an important readjustment in a person's life which changes its entire basis and accordingly affects the conduct of life and the emotional attitude towards it. If the Life Line ends at the first line across, the Line of Head, and joins it, we conclude that the person acts impulsively. If it ends lower down and does not join the Head Line (see fig. 67), he is not impulsive in his actions but deliberates and considers, a condition which developed to an extreme degree produces inhibitions. If the Life Line proceeds beyond the Head Line and almost reaches the second line across, the Heart Line (see fig. 69), we can assume that the person is hypersensitive and easily hurt, though he endeavours not to show his touchiness in front of others. He seeks to suppress it because he is not sufficiently conscious of the inner connectedness of his weakness, or tries to avoid the realisation of this knowledge. This kind of hyper-sensitiveness and vulnerability, however, exerts a much greater influence upon the soul than a touchiness which leads to spontaneous reaction and thus liberates and neutralises itself.

Small lines which branch off from the part of the Life Line facing the Mount of Venus and proceed in the direction of the lower part of the Mount of Venus mean that the vitality of the person has been impaired by illnesses (see fig. 70).

The division of the Life Line according to time can be carried out in the following manner: the first half of the entire line comprises the first twenty years of life. The second half is divided into two. The first half of this represents the twentieth up to the thirtieth year. The remaining part is divided once more and its first half assigned to the age between thirty and forty and the last section to the last period of life. (See fig. 71.)

As can be seen from the drawing, the first twenty years occupy the greater part of the Life Line, whereas the space allocated to the later years grows increasingly narrower and smaller. This apparently strange position is, however, in complete agreement with the psychological fact that the time conception of children and young people is more unlimited than that of older people. The child has very little conception of time and space, whereas from twenty onwards time becomes a factor increasingly to be considered, and with advancing age the conception and consciousness of time become as it were more compressed.

Two more features must be mentioned in connection with the Line of Life : the *Line of Prevention* and the *Line of Anxiety*. What I call the *Line of Prevention* (see fig. 72) is a line on the Mount of Venus which takes a parallel course to the Life Line nearer to the root of the thumb. It very often appears in the hands of people of a more or less weak constitution who unconsciously try to counteract this by preventive measures. This often results in a hypochondriac attitude, as such people when they have a slight cold are afraid of contracting pneumonia or some other serious illness and take elaborate precautions against blood-poisoning when they have pricked their finger : they are equally cautious in matters of hygiene and nutrition. These measures are, of course, not without advantage for the preservation of good health, but on the other hand consume such an amount of vital energy that the joy of life is impaired. The Line of Prevention may, however, indicate a type of man whose preventive measures are not concerned with his health but his material welfare. Such a person would arrange financial matters very carefully and economically and always plan things far ahead and with great deliberation.

The *Line of Anxiety*, as I named it, is a smaller or bigger stroke in the upper part of the Mount of Venus quite close to the Life Line (see fig. 66, line *k*). Fundamentally this line reveals fear of life, the fear to be unable to master life either materially or spiritually. It can also indicate constant anxiety for the well-being and life of the children or very close friends and relations. Sometimes it symbolises an extreme identification with the worries of others. I would recall to you that the upper part of the Mount of Venus represents sublimated impulses and instincts, whereas the lines on it indicate repressed energies. Thus it is perfectly logical that the Line of Anxiety should appear in the upper region of the Mount of Venus, for anxiety and fear are always the result of repressed instinctive energy.

The next line which attracts our attention is the *Line of Fate* (see fig. 66, line *b*). It can be recognised by its course, which starts at the wrist and proceeds towards the middle finger. Only a line running in the direction of the middle finger or ending at its base can be claimed as Line of Fate.

This Line reveals the capacity of a person to attune his inner life to the demands and conditions of his environment and his work. It shows the attitude and reaction of a person to experiences, impressions and events as they happen during his life, and also to what extent within the range of his abilities he manages to adapt himself to such events. Thus the Line of Fate is to a high degree the line of the *persona* and of adaptation to the influences of the environment. In most of the books on chirology and chiromancy this line is called the Line of Luck, and according to their static interpretation they consequently infer that it signifies luck and good fortune in the external sense of money, position, and success. One might indeed describe the ability of attuning one's inner life to the demands of the outside world and work as a happy gift, if one agrees that this is the necessary condition for the achievement of inner harmony and contentment, which is, perhaps, the only enduring and unassailable happiness. As the Line of Fate indicates the balance between the inner personality and the outside world, it is quite logical that it should run in the direction of the middle finger, which, as will be recalled, is the finger of mediation. To ascertain its significance and value for the personality we must observe its strength, its starting-point, its course and its termination.

As a rule, the Line of Fate begins at the wrist between the Mount of Venus and the Mount of the Moon and proceeds in a straight line towards the middle finger. The higher it ascends the greater the forcefulness and power of action of the personality, the greater the person's adaptability. Active and industrious people have a strong Line of Fate, and more passive, leisurely or lazy people a weak one. It must be pointed out, however, that not only the Line of Fate but also the thumb and the various features of the back of the hand must be taken into consideration. If the hand, for instance, is weak in expression and shape and the thumb weak as well, but the Line of Fate strong and straight, we can assume that the person has great ambitions and wants to play an important part but lacks the strength to assert himself, a position which often leads to a neurotic condition. In cases where the Line of Fate does not go beyond the first cross line but stops there, we can conclude that the life of the person is so exclusively guided and dominated by a one-sided line of thought or by a predominatingly intellectual and rational attitude, with a note of strong introversion, that the development of this person must come to an early standstill. This position prevents the person from adapting himself to his environment, which condition consequently produces symptoms of severe repression. Sometimes the Line of Fate stops abruptly at any given point of the hand, but is at once resumed and continues its course (see fig. 67). Such an interruption indicates a professional change or an alteration in the conduct of life or an inner readjustment, sometimes even a complete change of the style of life. A Line of Fate composed of numerous little lines and threads (see fig. 72) shows great uncertainty and undecidedness which produces continual professional and other changes in the conduct of life. This chirological feature is in most cases combined with a strong milieu attachment of the person, who constantly compares reality and the realities of life with the ideal image which he has made for himself of his father or mother. As such people in many instances were spoiled by their parents, nothing can fill their place and they cannot stay anywhere for long because they do not find the kind and indulgent treatment they are used to receiving from their loving parents.

The absence of the Line of Fate shows that there is fundamentally no possibility for the person to adapt himself socially. This absence of adaptability is often combined with a weak will which is indicated by a weak thumb. It is very often a lack of adaptability which, according to the type of hand, may produce criminal tendencies, and often the missing Line of Fate indicates impotence which, after all, is a failure to establish contact. I mention again that a weakly developed or altogether missing Line of Fate before the twentieth year does not permit of a final conclusion, as it may still appear or be strengthened. An important point in the interpretation of the absence of the Line of Fate is the fact whether it is missing in both hands or only in one. If it is missing in the right hand only it shows that the ancestors were badly adapted to their environment or were asocial in their attitude to others. Such a person may be influenced by this but is not necessarily afflicted with the same quality as far as his own character is concerned. If it appears in the right hand only, the person himself does not show the characteristics of this feature. The absence of the Line of Fate in both hands, however, suggests a hereditary anomaly and I have found this confirmed in the greater number of the serious psychopathic and criminal cases I have been studying. I have made the same observation in some cases of mental deficiency and idiocy. I should like to add that in the latter case the nature of the Head Line is of great importance.

The absence of the Line of Fate, however, does not necessarily indicate a poor intellectual capacity but fundamentally and chiefly the absence of the quality of social adaptability. I have, for instance, met several creative personalities, artists as well as scientists, who have accomplished great things in their domain though they had no Line of Fate. Their attitude, however, to their environment and the outside world in general could only be described as completely asocial.

In connection with my previous statement that the Line of Fate in order to be accepted as such must end at the base of the middle finger, I must point out that there is one exception to this rule : in some hands the Line of Fate does not run to the middle finger but turns to the first finger or index. If this is the case, ambition is so central and dominating a factor in the person and in his reactions to life that all else recedes into the background or is entirely obliterated. There can, of course, be no harmony in the mind of such a person, as everything is pressed into service for the attainment of one goal only, and a balance with the other parts of the personality cannot become effective. The possibilities of development of such people are very limited because they are too one-sidedly preoccupied with and interested in the realisation of an external goal. Their behaviour to the outside world is mostly unscrupulous, tyrannical and hard.

A Line of Fate which does not begin in the middle of the hand but at the Line of Life, or has very strong leanings towards it, indicates a late development or powerful suppression by a very dominating person in childhood resulting in a strongly inhibited personality with a self-assuredness that can be easily upset. Sometimes the Line of Fate which has developed late and emerges from the Line of Life may indicate that the person has spent the years of childhood quite unconsciously in an atmosphere of dreaminess and playfulness, or else that he has taken up the profession of his father.

A Line of Fate which starts directly at the Mount of the Moon (see fig. 68) and proceeds in a firm line right across the hand tells us that the person displays a tendency in his interpretation of and attitude to life directly opposed to those of the people of his milieu. Such persons usually show an early desire to assert their different attitude in an independent manner. Occasionally this feature points to a new profession which had never been taken up by any member of the family in previous generations. If the Line of Fate springs from the Mount of the Moon and there is no other line visible between the Lines of Fate and Life, we conclude that the milieu in which the person is growing up does not interfere with his independent tendency. Should there be, however, another line between the Lines of Fate and Life which takes a parallel course to the Line of Fate but comes to a much earlier standstill—it never goes beyond the Head Line—we can infer that the person finds within his milieu an impeding force or even opposition to his individual tendency. I have called this line which appears between the lines of Life and Fate—and which can be recognised by the fact that though parallel to the Line of Fate it ends much sooner—the *Line of Milieu* (see fig. 66, line *g*).

The Line of Milieu reveals that either the milieu follows a different tendency or is in opposition to the person, or that he has been subjected to a too severe and one-sided education and attitude to life based on conventions and traditions ; but it is also possible that the person is either in opposition to his milieu or too deeply attached to it. Whichever may be the case, the person is prevented from a free development of his personality and the unhampered display of its energies by disturbances or attractions which are expressed in his Line of Milieu. Certain parts of the personality which I have discussed in Chapter Six become bound up and thus unfree. The reasons for these restricting forces of the milieu will be found in the right hand and I refer once more to Chapter Six, where they are dealt with in detail. Moreover, we have to distinguish whether the influence of the milieu is recorded by one coherent line or several little lines and groupings of little lines (see fig. 73). If the Milieu Line is a complete line, it means that the person is continuously aware of the inhibiting and preventive influences coming from his milieu, if it is represented by arrangements or delicate little lines the person is unconscious of these influences. I have dealt with them in connection with the Mount of the Moon and recall that they may represent conflicts between the parents which, though not consciously experienced by the child, have had a decisive influence upon its unconscious emotional life, or repressions of one parent, or unconscious resistance of one parent against the child, or the emotional preference for a brother or sister by one parent or both. The essential and characteristic factor about these influences and impressions is that the person concerned is more or less unconscious of them and for that very reason their effect upon his development is even greater and more profound. They may have been experiences of traumatic shock and fear which were completely repressed and banished from the conscious mind. The small lines and groupings of lines occupying the space of the Line of Milieu which extends in a straight line between the Lines of Life and Fate I have called *Marks of Milieu.*

In the case of a hand showing a Line of Fate which springs from the Mount of the Moon and a Line of Milieu, the resulting difficulties described above nevertheless can become the motive power for personal development provided these difficulties are properly understood and the necessary adjustments made. If, however, the Line of Fate initially coincides with and later on emerges from the Line of Life and no Line of Milieu is to be seen, the attachment to the milieu is stronger than in the preceding case, because it means that the very roots of the personality are arrested by the milieu without the person being conscious of this. Thus with the absence of a contrast to or resistance against the milieu, the incentive for personal development cancels out.

A Line of Fate in the hand of a woman which has its origin in the Mount of the Moon indicates a note of masculinity, particularly if this is confirmed by a masculine appearance of the outside of the hand.

If the Line of Fate ends at the Line of Heart and the Line of Intuition in this period becomes more prominent, we can conclude that a turning point in the attitude to life and a change in the goal of life has been reached : the whole development has taken an inward trend, that is to say the goal of life as well as its direction are more guided by spiritual and personal factors than by external aspirations directed to the outside world.

To divide the Line of Fate into its time elements we proceed from the base of the hand to the point where it meets the Head Line, allotting this section to the period from the first to the thirtieth year, each third of this line representing ten years. The continuation of the Line of Fate between the Head Line and the Heart Line corresponds to the following ten years from thirty to forty. Half the distance from the Heart Line to the base of the fingers represents the next ten years, that is to fifty : the second half is halved again, which brings us to sixty, and, this process is repeated for the remaining portion of the line, the first half of each division always representing ten years. (See fig. 71.)

An important line which has a direct bearing upon the development of a person must be mentioned here though it is not a main line. It starts earlier than the Line of Life on the lower part of the Mount of Venus, cuts the Line of Life in its beginning and is either received by and joins any line, or terminates somewhere in the hand. I have very often come across this line in the many years of my chirological practice, and as my observations have taught me that the mothers of children who possess this line have had a decisive and palpable experience during the period of pregnancy which definitely influenced the psychical development of the child, I have called this Line the *Pre-natal Line* (see fig. 66, line *f*). In a great number of cases I have found this line in the hands of women whose mothers during pregnancy wished with great intensity to give birth to a boy. These girls developed definite masculine traits which did not exactly increase their happiness. Sometimes the mother had suffered a serious inner shock during pregnancy or had to go through a grave matrimonial conflict with the father of the expected child and later on the child seemed to be very much afraid of the father, though there was no apparent cause for the fear of the child in its parent's behaviour (see Appendix, plate 5). Sometimes the expectant mother went through a series of illnesses and lived in constant fear that the child might inherit a disposition to ill-health. A child born under these conditions may after birth become a normally healthy person but subject to a hypochondriac fear of illness.

We have learned that though the beginning of the Pre-natal Line is definitely located its end is not. Sometimes it is absorbed by the Line of Fate, the Line of Milieu (see Appendix, plate 5), the Line of Intuition, in fact by any line, or else stops at any point in the inner hand. The terminating point of the Pre-natal Line tells us in what way the trauma of the mother during the state of pregnancy has become effective for the owner of the hand. The importance and significance of the Pre-natal Line lies in the fact that it helps us to understand many otherwise inexplicable phenomena of the mind such as fear complexes, inferiority complexes and other inhibiting influences.

It is important that the interpreter of the Lines of Fate and Milieu should compare their relative strength. If, for instance, the Line of Milieu is stronger and more definitely outlined and not shorter than the Line of Fate, we can assume that the influence of the milieu and the attachment to it are so powerful that they hinder and bottle up the individual development of the personality. Should, however, the Line of Milieu, though more clearly cut than the Line of Fate, be much shorter than the latter, our conclusion is that in spite of the powerful influence of the milieu there is a definite possibility of its being counteracted and neutralised by psycho-therapeutic means and as a consequence for the personality to be freed and its development stimulated.

The first line across, which begins at the end of the Life Line below the first finger and proceeds right across the palm, is the *Head Line* (see fig. 66, line *c*). It reveals the kind and degree of a person's intellectual capacity, the state, disturbance and defects of the nervous system as well as head injuries and organic diseases of the brain. It should not run in a completely straight and rigid line across the hand but should possess a certain flexibility and be slightly curved.

The kind and degree of a person's intellectual capacity can be judged by the manner in which the Head Line ends and the exact spot where it stops. If the line is sharply outlined and ends below the second finger (see fig. 67) a sharp, cold intellect with a definitely rational and juridical trend is indicated.

A nimble intelligence, an intuitive mind with an artistic, creative tendency is suggested by a Head Line which ends below the third finger or between the third and fourth finger and is descending slightly (see fig. 68).

If it ends below the little finger (see fig. 69) the person has a definitely scientific mind and a good capacity for natural sciences, provided the Head Line is not completely straight but slightly inclined.

A Head Line which is so short that it reaches its end before the second finger shows a backward and weak intellect. I have also seen this type of Head Line in the hands of people suffering from organic brain diseases.

A Head Line which is split in two under the second finger

(see fig. 72), that is to say, terminates in a two-pronged fork, is an indication of a delicate nervous system usually overstrained by a premature and too intensive exertion of the brain. This Head Line will be found frequently in the hands of people who were intellectually precocious and accomplished intellectual feats which were far beyond their age. Their precocity favoured a too exclusive development of the intellect at the cost of other functions, thus undermining and weakening their nervous constitution in general. This kind of Head Line is thus, to a certain extent, a sign of degeneration, for the premature intensive activity of the brain and the resulting one-sided development of the intellect produce also a premature standstill of the general development, and very often a hypercritical mind—a rather negative virtue.

The Head Line which, however, ends in a fork below the third finger indicates a constructive, positive, critical intellect (see fig. 73).

A Head Line with a strong inclination towards the Mount of the Moon (see fig. 74) reveals a person who is easily depressed, unless it occurs in a clearly rectangular hand. As such a hand suggests a rather rational and sober-minded person the irrational phantastic element expressed by the Head Line descending towards the Mount of the Moon acts as a corrective and thus has a positive effect.

If one branch of the prematurely split Head Line goes to the Mount of the Moon (see fig. 70), the person is subject to serious depressions and a critical condition, as these fits of depression are not meeting with any resistance. Why a tendency to depression is expressed by an inclination of the Head Line to the Mount of the Moon finds its explanation in the fact that the Head Line, which is an expression of consciousness, drops into the unconscious, the unlimited region of phantasy, unchecked by the conscious mind. On the whole, this danger of letting oneself drop is less acute with a person the rectangular shape of whose hand indicates that his conceptions are rational and sober and based on the world of facts. If the Head Line conveying depression in a hand of mixed shape, and more so in a conical hand, is accompanied by such features as a weak thumb, a Line of Anxiety, a feeble Line of Fate, the tendency to depression is accordingly emphasised and increased. Sometimes, however, it happens that a person is temporarily subject to depression and yet the above features do not appear in his hand. The explanation is that his melancholic disposition is not constitutional but caused by a lack of inner balance and self-assuredness, a feeling of guilt and all kinds of nervous disturbances, or to formulate it in a more general way a lack of consciousness regarding the emotional side of the personality. These symptoms of depression can be overcome to a great extent by a development of the unconscious forces and by becoming conscious of these problems.

A Head Line which runs in a broad line right across the hand (see fig. 75) shows a very undifferentiated, rigid intellect, wholly concentrated on the material side of life. Such a Head Line seems to cut the hand into two parts, separating the lower section, representing the earthly sphere, from the upper section, the region of the spiritual. It suggests that flexible adaptability is lacking, and emotion and reason very often pursue their own separate paths. Such a Head Line which runs right across the hand is often a sign also of insincerity and a mendacious disposition.

If the Head Line is very thin, a weak nervous constitution and a poor memory are indicated, but if it is strong and slightly chainlike (see fig. 76), the powers of combination and association and the memory are good.

There are hands with actually two Head Lines—that is to say, not a split line but two separate lines, one of which may start at the Life Line and the second away from it, assuming a different shape, often pursuing a different course (see fig. 77). This suggests a certain duality of personality which gives it a double-faced character. The one part of the personality, as it were, does not know what the other doeth. Such people may in course of a conversation express a negative opinion on something although only an hour earlier they made a positive statement about the same matter to somebody else, without being aware of a discrepancy in their judgment. With such people one can never be sure what their disposition will be and what they really think of a person or matter, as their attitude and their opinions seem to change constantly and even contradict each other. Though, as a rule, they are highly intelligent people, their two-faced personality makes them unreliable and gives them the stamp of insincerity. The double Head Line sometimes appears in the hands of people who actually lead a double life, a Jekyll and Hyde existence, leading the life of an ordinary citizen by day and that of an adventurer by night. The significance of a double Head Line is thus generally speaking a periodical split of a part of the personality.

A branch of the Head Line going upwards, that is to the Heart Line, and another leading downwards to the Mount of the Moon is a sign that the person has a disposition for pathological melancholia (see fig. 80).

An interruption of the Head Line at any point suggests that the person must have suffered from nervous trouble at that particular time (see fig. 81).

If the Head Line descends in a steep drop to the Mount of the Moon (see fig. 82), that is to say not with a slow movement but unexpectedly and suddenly, this is a sign that the person is subject to serious, unexpected and abrupt fits of depression and also, according to the type of hand, of brutal callousness, especially if the Line of Heart is straight.

We have found that the cleaving of the Head Line below the third finger means constructive criticism. Now if one prong of the fork clearly proceeds in the direction of the Mount of Mars and the other to the Mount of the Moon (see fig. 83) we may infer that the person subjects himself to severe self-criticism and is inclined to doubt his own personality or even negate it. Towards others, these people mostly adopt an attitude of benevolent criticism.

A kind of hump on the Head Line below the second finger (see fig. 84) suggests insincerity and a mendacious disposition.

A Head Line can be called good if it proceeds in a supple movement and ends slightly bent below the third finger or a short distance behind it, becoming fainter in outline towards its end (see fig. 68). A Head Line which takes a completely straight course and maintains the same intensity of outline throughout must be considered as bad (see fig. 75).

The division of the Head Line according to time, that is to say the time of events and impressions connected with and reflected in the Head Line, can be established in the following manner : one draws a vertical line (in the left hand) from the right-hand side of the index finger to the Head Line and the point where the two lines intersect signifies the tenth year. Such a line drawn from the left of the middle finger yields the twentieth year, from the right of the middle finger the thirtieth, from the left of the third finger the fortieth, from the right of the third finger the fiftieth and from the left of the little finger the sixtieth year of life (see fig. 71).

The most striking fact in this connection is the big span occupied by the periods between the twentieth and thirtieth, and the fortieth and fiftieth years. The explanation will be found in the importance of these two periods, which are decisive in the development of the intellect. In the first period the intellect is sharpened and developed in battling with the scientific knowledge so far accumulated, a fact which is borne out by a comparison with the corresponding periods in the lives of all scientists. The second period is that of productive mental activity in which the acquired and assimilated knowledge is creatively applied in accordance with the individual experiences of the scientist.

The *Heart Line* starts below the little finger and should end in a fine curve between the first and the second fingers (see fig. 66, line *d*). It gives us insight into the emotional sphere and reveals dispositions for organic as well as nervous illnesses of the heart and disturbances of the secretory system. Thus in contrast to the Head Line it represents the unconscious part of the personality. If it ends in a gentle curve between the first and second fingers we can conclude that the person is warm-hearted, generous and emotionally responsive (see fig. 68).

A Heart Line which ends earlier, for instance before the second finger (see fig. 70) indicates an emotionally cold and perhaps also calculative type with an organically small heart.

A Heart Line which ends with a fine sweeping movement on the Mount of Jupiter shows the definite prominence of ambition in a person's emotional life, which, however, need not rule out benevolence and warmheartedness (see fig. 69).

If the Line ends on the Mount of Jupiter but takes a completely straight course without the slightest curve we may conclude that such a person is without any consideration for others and pursues his goal with cold determination and callousness, and that his emotional life is dominated by his ambitions and his craving for power (see fig. 85). Should several smaller branches and frayed lines in the direction of the Head Line appear at the end of the Heart Line (see fig. 74) the person is inclined strongly to waver between emotion and reason.

If branching off from the end of the Heart Line a line leads to the beginning of the Head Line (see fig. 79), the intellectual processes strongly influence the emotions. This feature, which can be found in many hands is not necessarily a negative trait, for it may be an indication that the person is compelled to control or disregard his emotions or has to deny himself many things for the sake of the general development of his personality. The control of the emotions often entails suffering which, if properly exerted, stimulates and furthers the development of the personality more than their spontaneous gratification. The same feature, however, must be interpreted as a negative trait if the expression of the hand as a whole and other signs as well show a materialistic tendency. In the reverse case, that is to say, if a branch leads from the end of the Head Line to the beginning of the Heart Line which, as a matter of fact, is less common than the arrangement described above (see fig. 78), we conclude that the intellect is strongly influenced by the emotions.

A Heart Line which is well curved but sends an arm to the Mount of Jupiter or the Head Line suggests goodnaturedness with a purpose, coupled with a strong desire for recognition. Such people are only kindhearted and prepared to help when they are sure of recognition ; they must always have a definite purpose in mind and achieve something " real " with their help.

As with the Head Line, a Heart Line which goes right across the hand in a completely straight line has a barrier effect. Because of a lack of differentiation in the emotional sphere some function or other has either been eliminated or at least been greatly weakened. People who are in the habit of making quite unexpected and unmotivated decisions as well as cold and calculative persons have this line.

If the Heart Line goes straight across the hand and the Head Line springs from it below the second finger and proceeds in a short and sharp line, we conclude that the person is emotionally undifferentiated and undeveloped, with a predominating intellect. As a rule, such people are very intelligent but selfish, calculative and without emotional finesse (see fig. 85).

Should the Heart Line and the Head Line proceed as one line we must ascertain which section of the line is more strongly and purposefully designed. If the section which starts below the little finger is stronger (see fig. 86) we can assume that the person is without inhibitions, erratic, strongly impelled by his instincts, and as a rule incautious, which exposes him to the danger of suffering bodily injuries and of meeting with accidents. If, however, the other part which begins below the first finger is stronger, we conclude that the person is calculative, cold, unemotional and egoistic (see fig. 87).

A Heart Line of a definitely chainlike texture (see fig. 72) suggests disturbances of internal secretion. Organic heart trouble is indicated by interruptions of the Heart Line or the formation of many islands (see fig. 73). If the Heart Line is unusually red we assume that the person is very excitable and that something is wrong with the vascular functions.

A sensitive gall or liver produces a particularly dark colouring of the initial section of the Heart Line. Sometimes another line in the hand which will be discussed later emphasises this gall or liver disposition.

Many small descending lines which occupy and intersect the Heart Line reveal a strong desire for life experiences of an ever-changing variety which is often connected with an undeveloped and poor emotional capacity, a phenomenon we met before and discussed in connection with the third finger (see fig. 77).

With the Heart Line the division of time is found by the reversed process adopted in the case of the Head Line. Thus at the point where a vertical line drawn from the left side of the little finger of the left hand intersects the Heart Line, we locate the tenth year. The right side of the third finger yields the twentieth year, and so forth in sections of ten years as in the case of the Head Line. These sections provide us with a fairly accurate means of calculating at what age certain experiences which left their trace on the Heart Line actually happened (see fig. 71).

Normally there should be an interconnection between the Heart Line and the Head Line represented by a number of smaller connecting lines in the space between the two main lines (see fig. 76). They record that the impressions and phenomena of life generally, and those of the environment, are simultaneously received and homogeneously assimilated by the emotional as well as by the intellectual part of our personality. If these lines do not appear we must conclude that the emotions and the intellect go their own separate way and that the reactions of such persons to their environment are correspondingly divided. Thus we find in this feature another barring effect.

In the space between the Heart Line and the base of the fingers we often find, apart from the termination of the main lines, a number of cross-shaped lines which, if they are very numerous, indicate that the person is very emotional, highly excitable and lacking in self-control (see fig. 81).

The four main lines form as it were the fundamental structure of the personality, and their course, intensity and direction must always provide the essential basis for the interpretation of the remaining features of the inner hand, as the other lines, which will be the subject of the following pages, are not part of the constitutional structure of the hand.

The line which apart from the main lines is, on the whole, most frequently found in the hand is the *Line of Intuition*, for after all intuition is an integral part of the personality. As I have already dealt with the functioning of intuition in Chapter Four, I need only add that the Line of Intuition as well as the other lines connected with it show the effect of the functioning of intuition upon the mind. In the course of many years of observation, I have found that there are various kinds of Lines of Intuition and I have classified them according to the various types of intuition they represent: creative-productive intuition, combinative intuition which is based on experience, and, thirdly, imaginative intuition.

In order to decide which kind of Line of Intuition it is, we have to study the beginning and the end of the line. The termination of all Lines of Intuition is the third finger or the space between the third and fourth fingers. A Line of Intuition which ends on the third finger suggests an artistically directed intuition, particularly for music, and also, according to the type of the hand, possibilities of spiritual development (see fig. 66, line *l*). If the line comes to an end between the third and fourth fingers, the left side of the little finger being the limit (see fig. 66, line *i*), an intuitive gift for technical and practical things and matters of organisation is indicated.

The *Productive Line of Intuition* begins at the Life Line. The person who has it possesses the power not only to appreciate but also to shape and creatively to form, an intuitive power which, as I expounded in Chapter Four, is independent of experience and is present in man from his birth. The intensity and course of the line show the degree to which it is effective. The significance of this line lies in the fact that people in whose hand it appears—provided it is well shaped and strong—are able not only to see and to perceive things as they appear in the world of reality but also in their partly unconscious, inner reality.

The *Combinative Line of Intuition* which has its roots in experience—I should, therefore, like to call it the Intuitive Line of Extraversion—starts at the Line of Fate. As indicated by its name, it reveals the ability of appreciative assimilation derived from experience and thus supports productive intuition. A physician, for instance, who examines human bodies every day does not register all his experiences in his brain as in a file. They become part of his unconscious mind and with every fresh case he can draw upon this unconscious store of experiences for diagnostic and therapeutic purposes without having to muster his entire creative power each time.

The third line, which I have called the *Imaginative Line of Intuition*, springs from the Mount of the Moon and usually ends on the third finger (see fig. 73). The intuition it represents comes from the collective unconscious, from the imagination. It does not perceive the phenomenal world like the productive intuition in its actual but in its " unreal " reality, and is concerned with the possibilities of the inner development. One will find this particular Line of Intuition mainly in the hands of clairvoyants who always see beyond reality, of painters who do not paint reality photographically but the inner reality as they see it, and of good story-tellers who have to rely on creative imagination. Imaginative intuition, in order to have a positive effect on the mind as a whole, needs the support of good combinative and self-critical qualities, for in cases where these qualities are lacking there is a danger that the subjective imagination loses itself in the irrational and projects its own unconscious attitude into the matter it is dealing with. One can often find this line without those corrective lines in the hands of clairvoyant people who, because of this lack of self-criticism, are unable to see their own weaknesses and shortcomings and are thus inclined to transfer them and to project them into their prophecies, a fact which explains why they are always predicting negative and bad things.

A very strong Line of Intuition may, however, have a negative significance if the Line of Fate happens to be very weak or altogether absent. Such a constellation indicates that the persons in whose hands it appears have no sense of reality, no adaptability and are, therefore, inclined to live in a fantastic world of their own, to become cranks and asocial idlers.

There is, however, a fourth Line of Intuition which is chiefly found in the hands of women and springs from the Mount of Mars. Persons who have it seek friction, disputes and quarrels in order to feel and display their own strength. The reason why it occurs more often in the hands of women may be found in the fact that men on the whole encounter sufficient struggle and friction in their professional life and work and, therefore, have little desire to carry them into the private or domestic sphere. Women who have no such outlet and whose life is generally more uneventful and monotonous are, according to their disposition, more or less inclined to break the monotony by a bit of domestic strife and personal quarrels. This Line, of course, occurs only if there is a Mount of Mars.

A Line which does not terminate on the second or third finger but between them and, therefore, cannot be looked upon as Line of Fate or Line of Intuition, reveals a tendency for religious activity within a particular movement.

The preceding observations show us distinctly that the Lines of Intuition represent the reactions of the unconscious mind, whereas the Line of Fate is a record of the conscious mind of the *persona*. We can assume that the intensity of the Line of Intuition is not so much conditioned by the degree of intuition actually displayed as by the instinctive and unconscious use made of it within the personality. That is why the Line of Intuition must be complemented by the Line of Fate.

Though intuition is originally an unconscious force, it is possible, and even necessary, to make it conscious. It can be observed that with an intensive, unconscious intuition the sense perception often seems inferior. Now if this inferior sense perception is developed, the power of observation is improved, which in its turn makes the intuition more conscious, since by this a stronger control of the perception by means of intuition is achieved.

The process of making intuitive perception more conscious which increases the effectiveness of the intuitive power in the entire psyche may have a beneficial effect on the personality as far as self-assuredness is concerned. One often finds a disturbed or entirely repressed intuition in people who are emotionally inhibited or whose self-assuredness is impaired because of a milieu-fixation, a state of mind which makes them transfer the centre of their reactions to the intellectual sphere. As their self-assuredness has been upset they have lost confidence in their emotional and intuitive powers and, therefore, force themselves to see and to judge everything rationally, an attitude which, however, does not make them free or happy. Often these disturbances of self-assuredness caused by milieu-fixation, or opposition to and hatred of the milieu, are combined with feelings of guilt. Such feelings produced by the above-mentioned reasons are almost exclusively due to the fact that the person failed to face and to settle negative experiences in a positive manner. These feelings of guilt also exert a disturbing and impeding effect upon the intuition. The strangling of intuition is chirologically expressed by the fact that the Finger of Intellect is too long and usually inclined towards the finger of intuition.

If by the process of a conscious investigation and clarification of the problem under discussion a consolidation of self-assuredness and self-confidence has been achieved, the intuition also has become more conscious, which renders it more effective and more valuable for the mind as a whole.

CHAPTER NINE

SUBSIDIARY LINES AND VARIOUS OTHER FEATURES OF THE INNER HAND

We have seen that the four main lines represent the constitutional structure of the personality as far as its reactions are concerned. Thus it is very important for the interpretation of the reactions and the attitude of a person as far as his personal individual experiences and impressions are concerned to observe how these main lines are touched, intersected or interrupted. There is also a multitude of other features of which only the most important and most frequent can be discussed, since their combination of form and the expression of the hand, fingers and lines are so complex that their observation and interpretation must be referred to the domain of practical study.

The *Girdle of Venus* either connects the little finger with the index or with the Mount of Jupiter (see fig. 69), thus linking ambition and self-assertion with the sex urge ; it shows that some peculiarity in the sexual life makes this the centre and focus of the personal life. According to the type of the hand, the Mount of Venus and the little finger, this peculiarity can suggest hypersensitiveness, an overdeveloped aesthetic attitude in sexual matters which will produce disturbances and inhibitions, but this peculiarity may also manifest itself as an extreme sexual desire without restraint or inhibition (see Appendix, plate 7). The connection of the Mount of Jupiter with that of Mercury reveals that personal ambitiousness and craving for self-assertion try to find satisfaction in the sexual sphere by acquiring and possessing as many partners as possible, which in most cases is an over-compensation for a strong inferiority feeling. This position can also be described as a transfer of the active and extraverted craving for power to the sexual sphere. Sometimes several smaller Girdles of Venus appear (see fig. 74), usually in the hands of highly strung, neurotic and hyper-sensitive persons, and when combined with other signs are almost invariably found in the hands of homosexuals.

The *Ring of Saturn* (see fig. 73), however, connecting the fingers of Mercury and Saturn or Apollo and Saturn, shows the tendency to sublimate the sexual urge, to spiritualise, to introvert it and thus forms a contrast to the extraversion expressed by the Girdle of Venus. I have described this phenomenon, which is particularly often found in the hands of women, as " vanquished resignation ". These women, who are usually plain but intelligent, have ceased to suffer because of their plainness and are thus able freely to develop their personality and individuality. The effect of the " vanquished resignation " is mostly beneficial, for the expression of spirituality and intellectual refinement makes the face attractive.

The *Poison Line* starts at the lower part of the Life Line and proceeds in a straight line to the lower section of the Mount of the Moon (see fig. 66, line *n*). I have named it Poison Line because people who have it feel an instinctive aversion to all kinds of poisons, such as allopathic medicaments, alcohol, tobacco, etc., and at the same time are very sensitive to the effect of these and other poisons. It may well happen that such people strongly react to quite small doses of allopathic remedies. An injection, for instance, by the dentist may cause them to faint, or certain medicines will produce fever or a rash. This line is therefore important to doctors for estimating the dose of a drug which will do no harm to the constitution of the patient. This is particularly important because up till now there is no reliable test to gauge the sensitiveness to poisons, a fact which has been pointed out to me even by homoeopathists, who use only minute doses. People who have this line are altogether instinctively inclined to lead a natural life and to observe a natural kind of hygiene. They are in favour of a vegetarian diet and strongly dislike and disapprove of surgical operations of any kind. If this line is not quite straight and interrupted at any point of its course, or provided with an island, we can assume that at some time in the past a medical drug has been administered which has done harm to the entire organism, for, as we know, people with a Poison Line cannot very well assimilate poisons or counteract their harmful effect. I want to stress once more that such people are by no means drug-addicts, but on the contrary show a pronounced aversion to all kinds of drugs and poisons.

This is confirmed by the fact that drug-addicts never have a Poison Line but a large Mount of the Moon without any lineation, combined in most cases with a white hand, a delicate back and long fingers.

Very frequently the Poison Line will be found in the hands of highly intuitive persons, and there seems to be a certain connection between this line and the development of the intuition, for if they disregard the warning of their Poison Line which tells them not to smoke, drink and take drugs, their intuitive powers are impeded. Yet it is true that they can increase their intuitive capacity by taking heed. The explanation probably lies in the fact that in the long run the taking of narcotics deteriorates the original powers of man.

The *Gastro-enteric Line* normally springs from the Life Line and in some rare cases also from the Line of Fate. It proceeds in a straight line towards the little finger and terminates—unlike the Line of Intuition, which shows technical ability and ends between the fingers of Mercury and Apollo—directly below the little finger. If it runs in a straight and pale line over the palm, the person has a disposition to gastric and enteric illnesses (see fig. 72).

Should the line not be continuous but divided into two separate parts (see fig. 69) and assume a darker colour than all the other lines, a disposition to gall and liver trouble is indicated. If, however, the line is severed into little bits (see fig. 68), this indicates that the person is inclined to suffer from illnesses of the kidney or the bladder.

Sometimes, instead of a whole line only a branch springs forth, not from the Line of Life but from some other line as, for instance, the Line of Intuition, and goes towards the little finger. This suggests a nervous disturbance of the enteric function, which often afflicts people who are suffering from stage fright or similar states of nervousness. It is not difficult to understand that the strong, and often too strong, exertion and concentration of the intuitive powers combined with suspense and fear should upset the enteric function.

A different interpretation suggests itself if the line coming from the Line of Life terminates on a prominent Mount of Mercury. In this case the line has no gastro-enteric bearing, but is merely an emphasis and strengthening of the quality of the Mount of Mercury, provided, of course, that there is such a Mount in the hand.

A line starting out from the Life Line and following the direction towards the little finger and stopping, without reaching its goal, in the middle of the palm (see fig. 66, line *o*), reveals a disposition to infectious diseases or else that the person has suffered from them in the past.

An *Island* is an interruption in a line, constituting an unoccupied space, a rupture (see fig. 73), and indicates a disruption or disturbance of the function or process concerned. An Island in the Life Line, for instance, shows that the life urge and the forces of vitality in the period concerned were impeded by illnesses, or sometimes by imprisonment. Islands in the Line of Fate are a sign that disturbances have occurred suddenly in connection with the work or the general external circumstances of life. In the Head Line they indicate nervous disturbances caused by serious shock or trauma-like experiences, or sometimes by head injuries. In the Heart Line they are the expression of emotional disturbances or injuries.

Squares (see fig. 73) reveal temporary impediments and difficulties which, according to the line where they occur, are mostly more or less connected with the influence of the environment. Thus a Square at a point of the Line of Fate and at the corresponding point of the Life Line can be an indication that during that period the person was not in possession of his freedom. Only Squares which are formed by three independent lines and a main line, or four independent lines, and not by two or three main lines, can be regarded as genuine Squares, for the crossing of main lines would inevitably lead to the formation of such. Moreover we must distinguish between large and small Squares, because of their different psychological significance. Large Squares on or between the lines usually represent experiences of which the person has no precise and definite recollection, whereas the small Squares stand for very clearly conceived experiences and events of which the person is definitely conscious. In most cases large Squares comprise a greater space of time, and their effect is more difficult to assimilate because it is less conscious. This renders them psychologically more important than the smaller ones, for the effect of the unconscious on the attitude of a person is greater than the effect of conscious experiences.

According to the line with which they are bound up the Squares show the reactions of the personality which is in keeping with the character of that particular line. If Squares appear on the lower part of the Mount of Venus and are linked with the Life Line, we can interpret this as a premature traumatic shock in the sexual sphere caused, for instance, by rape.

Squares on the Mount of Jupiter (see fig. 73) reveal that extreme ambitiousness and a violent desire for domination have led to great difficulties in all matters concerning the relationship with and adaptation to the environment. Sometimes these Squares point to a difficulty within the family, particularly if there is a strong Line of Milieu.

The significance of Squares also differs according to the type of hand. A few Squares in a scantily lined hand, for instance, are of more importance than a much greater number in a hand with an abundance of lines. The explanation will be found in the fact that people with many lines in their hands live more eventful lives and thus, because of their experiences and the pattern of their characters, get used at an early age to the struggle against difficulties and odds. Consequently their power of resistance is greater and more resilient, and thus a comparatively greater number of difficulties, represented by a greater number of Squares, would affect them less than fewer difficulties would affect persons whose hands show only a few Squares.

Whereas a Square is an indication of external or internal difficulties which, according to their kind, had been developing more or less quickly but gradually, an Island represents an event or happening in a person's life which occurred suddenly with elemental force and often without previous warning.

Triangles (see fig. 73) reveal progressive phases of development through inner realisation. Thus, for instance, if Triangles are connected with the Life Line we can assume that illness has given the person an opportunity for deep thought and contemplation and that this period of inner realisation has furthered his or her development. A development of the inner self can also be brought about by difficulties in the emotional sphere, and the result of this inner struggle is chirologically expressed by a Triangle on the Heart Line below the third finger. A Triangle which appears at the end of the Line of Intuition below the third finger shows an artistically gifted person with a strong pedagogical ability.

Lines which start on the Mount of Venus and cease at any point in the hand are *Lines of Experience* (see fig. 70). If they proceed right across the hand intersecting the Lines of Fate, Head and Heart (see fig. 76), we conclude that the experience was very decisive and complete and shared with a partner. Experiences with brothers and sisters, with children, or experiences of a more general nature, however, are indicated if the line starts on the upper section of the Mount of Venus (see fig. 74). Should the lines springing from the Mount of Venus run directly to the Mount of Mars, we know that the experience was of a negative kind. If the line leads to the Mount of Mars but springs directly from the Life Line instead of the Mount of Venus, we assume an experience connected with severe illness, death or loss of a person.

In conclusion, I should like to point out that all lines between the Lines of Life and Fate are connected with influences of milieu and environment. The lines in the space between the Line of Fate and the upper section of the Mount of the Moon and the Mount of Mars refer to influences of the outside world. The influences of one's own unconscious mind are recorded by lines ascending from the Mount of the Moon.

CHAPTER TEN

PRACTICAL NOTES

I suppose the reader has realised by now the truth of my previous statement that chirology cannot be learnt from books alone. Another reason why practical work is essential lies in the impossibility adequately to study the back of the hand and the fingers, their expression, colour and proportions by means of illustrations, because these chirological features cannot be reproduced with sufficient accuracy and naturalness. Moreover, the possibilities of combination are so vast that they could not be dealt with exhaustively in a book of average size. Besides, experience tells us that books containing an abundance of detailed information are apt to confuse the student and to make him uncertain. Thus it has been my endeavour in this book to confine myself to the most essential chirological features and traits, but I must repeat once more that in the study of psycho-chirology practical work and experience are of the utmost importance, particularly for the beginner.

Here a word must be said about the suitability of a person for the chirological profession, for in quite a number of books it is claimed that everybody is capable of learning chirology. This, however, is not the case, for apart from the necessary psychological training the chirologist must be intuitively gifted, possess powers of combination and the degree of intelligence that is expected of a student of medicine or applied psychology. Apart from these qualities the student of chirology must have patience and staying power. In the classes and seminaries I have been conducting for the last ten years, I have often found that students after about the seventh lesson begin to show signs of impatience when they realise that their analysis of a hand yields less comprehensive results than mine, which is based on many years of experience. These people are apt to forget that the study of the hand implies the study of the entire range of human nature, its functions, reactions, hereditary traits and relationship to the milieu, that its phenomena in their interconnectedness are so complex and ramified that a reasonable amount of time is needed to get to know them and an even longer period to be able to apply them correctly. On the other hand, my experience as a teacher has also taught me that, provided these conditions are fulfilled, chirology can be mastered, theoretically as well as practically.

In addition to my observations expressed in the second chapter about the personality of the chirologist and his attitude to his work, I must mention one more fundamental condition : love of mankind. To be a true helper he must try to discover in a person who turns to him for advice and help those traits which are capable of development ; he must treat him with sympathetic concern, yet objectively, for he should not only point out positive but also negative qualities. Personal tact is, of course, a most essential virtue of the chirologist, because much depends on the manner in which he formulates his findings.

The great advantage of the psycho-chirological method as compared with nearly all other psychological systems lies in the fact that the person seeking advice need not speak and in the beginning is not even called upon to answer questions ; he is told everything. This direct method of ascertaining the difficulties, problems and depressing experiences of a person creates a suggestive atmosphere of confidence and induces him to talk, thus considerably easing his inner tension and constraint. Very often the inner impossibility to talk constitutes the chief inhibition of many psychologically complicated and suffering human beings. A further result of this atmosphere of confidence is the readiness of the person to accept advice and guidance. Quite frequently the clear and unequivocal pointing out of a talent or a positive quality strengthens the poor self-confidence of a person or reveals to him for the first time that he actually does possess certain positive qualities.

One of the things which has to be pointed out to people who are battling with inner difficulties is their tendency to take themselves too seriously. Very often they are convinced that the experiences and sufferings they have to go through are quite unusual and unique. This attitude reveals a lack of collective consciousness and an extreme degree of ego-consciousness. The chirologist has to rectify this attitude by convincing such people that many of their fellow beings are afflicted with the same or similar problems and troubles, that they are over-dramatising their individual fate. They should take their problems seriously, but they should not invest them with an exaggerated importance. These two reactions differ from each other because persons who take their experiences seriously do not merely consider their cause but also the final goal, the destination towards which they point. Those, however, who are inclined to take their experiences too seriously, thus exaggerating their importance, are induced to contemplate only their cause and the temporary situation they have created, without realising their final purpose and true significance, and their inner conflicts will continue to dominate their lives. They must learn to understand that a true sense of proportion provides a possibility to digest and master one's difficulties. The occupation with oneself is unwholesome if it is solely concerned with the *persona*, that part of the ego which is merely related to the outside world. The right way of occupying oneself with one's ego consists in the searching of one's soul, the exploration of one's mind and its innermost recesses as revealed in the states of dream, contemplation and meditation. In these states of mind not only the personal reactions are experienced but also those which come from the collective emotional sources common to all human beings. By this process of transmuting collective inner experiences from the unconscious to the conscious sphere, the individual mind becomes reunited with the community of mankind, whereas exclusive preoccupation with the personal-individual part of the self can only produce a state of isolation. Overstressed ego-consciousness conveys a lack of inner humility and the inability to acquire the right attitude to the afflictions and difficulties of one's life. The person who seeks advice must also learn to realise that there is no true development without difficulties, that negative experiences are essential because they build up our inner strength and establish our independence. All opposition, if offered in a constructive and positive spirit, generates power, but this power is only produced when we are prepared to accept the rough with the smooth, that is to say, when we are ready to accept that which is negative and painful in our lives as a necessary part of it. The unduly ego-conscious person does the opposite. He has the feeling that he is being treated unfairly by a destiny which has made his life so full of difficulties, and consequently he is always embittered and has a grudge against all harmonious people. These negative reactions are bound to have a negative effect upon the development of his personality. He refuses to accept himself as he is, and the suffering imposed upon him, not knowing that in order to apply one's powers freely and constructively it is a primary necessity to accept oneself as one is by disposition. The constant wish to be different shows a lack of humility and religious sense, because it amounts to a protest against the creative powers and reveals the arrogance of a man who presumes he can improve the work of creation. In sensitive people this attitude is bound to produce inferiority feelings.

On the other hand, such inferiority feelings can be used to further the inner development, provided the person understands the cause of these feelings as well as their effects. This acceptance of oneself and the realisation of one's weaknesses does not mean that we should indulge in them and give in to them all the time ; on the contrary, this positive attitude gives us the necessary freedom to go against them.

The first suffering that afflicts the young person is caused by his struggle with external reality. In the paradise of childhood the little child lives completely in the realm of his inner reality, trying to interpret the phenomena of everyday life as miraculous happenings and to merge them in his own imaginary world. Gradually the child awakens and the necessity to adapt himself compels him to accept and to recognise the phenomena of the external world of realities as such. The way in which this experience is dealt with and assimilated is decisive for his whole future development. Every person who wishes to return, fully conscious and matured, to his inalienable paradise, that is to say, to unhampered activity and full realisation within the domain of one's inner life, must go through this process of reconciling the external with the inner reality. Inner realisation cannot be obtained by ignoring outside reality. If scorned and suppressed, external reality takes its revenge by throwing stumbling-blocks into the path of development, by assuming domination. Should the individual try to escape even deeper into his inner self, he lays himself open to all sorts of inner conflicts and clashes with the outside world. If one duly respects the given phenomena and laws of the external world and accordingly adjusts one's life, man is no longer the slave and dupe of externals but is free to express his inner reality. Unfortunately most people tenaciously and anxiously cling to their childhood paradise and refuse to take cognisance of the given conditions of the reality of their environment, thus unwittingly jeopardising their attainment of the more enduring paradise of true inner realisation.

The first object involved in the struggle with reality is usually found to be the parents, and the first blow is the realisation that they are not as perfect as the child thought them to be or, in other words, that they do not live up to the ideal parent image created in childhood. The clinging to this conception of the perfect parent is connected with the pleasurable attitude of the child who sees and wants to see an ideal in its parents, thus unconsciously creating a basis for its own self-assuredness. It is obvious that this basis cannot be a genuine and lasting foundation, for parents, like all human beings, cannot be perfect. This attitude of the young person to his parents is one of the causes of a milieu-fixation. The proper and normal release from the parents can only be effected by the attainment of an objective attitude and by the realisation that parents are just as imperfect as other people. The result of this discovery is that the young person tries to stand on his own feet and to achieve independence, a step which greatly influences his attitude to the entire world around him.

It is often necessary to enlighten people about their so-called inferiority feelings, which, by the way, are mostly afflicting those who have not the strength of mind to admit their actual points of inferiority to themselves. The fact that inferior traits of character and mind occur is no less surprising than the fact that human beings have organic inferiorities. There are two groups of people who suffer from inferiority feelings : those who are aware of their actual inferiorities but do not want them to be true and, therefore, fight against them, and those who do not even admit that they have them. The people of the first group are suffering from intense feelings of displeasure, discontent, moodiness and fits of depression, but they are comparatively objective in their judgment of their environment and not intolerant to others. Those of the second category are inflicted even more intensely by the above unpleasurable feelings, but they are intolerant to others because they can only see their negative traits and are particularly inclined to criticise their own personal shortcomings in others ; they are, in fact, projecting their own inferiorities into the outside world. Those, however, who are aware of their inferiorities do not suffer from them in such a negative way, because they have the possibility to combat and to remove them. Thus it is most essential to induce those people to realise their own shortcomings and to accept them as given facts.

The discussion of these problems and the advice given on these lines can have a clarifying effect and be of considerable help to the persons concerned by furnishing them with the possibility to overcome their difficulties.

Often, however, the causes of the difficulties and problems are so deeply submerged in the unconscious that the case has to be referred to an analyst.[1] Here, too, psycho-chirology can achieve favourable results because of its method, which does not depend on questioning the patient. By explaining to persons the nature and inner connectedness of their problems and difficulties, I have often succeeded in one meeting in convincing them of the necessity of an analytic treatment, although many of them had hitherto refused to take such a course in spite of the advice of friends and doctors. Though I must mention the fact that not everybody is suited for analytic treatment, a point which will be explained in the following paragraph, it would be rash and show a biased mind to condemn analysis on principle. This hostile attitude, which is not confined to any particular group of people, is usually assumed by persons who have never gone through an analytic treatment themselves, and I maintain that it is impossible correctly and objectively to judge the analytic method without having experienced it. Frequently the whole method is condemned root and branch because these severe judges have heard of a few unsuccessful cases. The same people would not dream of anathematising surgery because a certain number of operations are fatal. Besides, the value of analytic psychology and its importance for the treatment of severe neurotic and nervous diseases, as well as for the inner development of man, are now so firmly established that there is no need for me to take up the cudgels in defence.

People who are not suitable for analytic treatment are the spiritually primitive, the completely unintelligent, the purely materialist and the mendacious types. Another type unfit for analysis, the person who seems to be completely submerged in his unconscious mind, can be chirologically recognised by the fact that the fingers compared with the back of the hand are unusually long and often also very thin, that all fingers are inclined towards the sphere of the unconscious or else the third and fourth fingers are set apart from the other fingers and tend

[1] Whenever I make reference to analysis I mean Professor C. G. Jung's system of analytic psychology, which I know, theoretically as well as empirically.

towards the unconscious, that the Mount of the Moon is covered with a maze of lines which lead into the void, that is to say, do not link up with any of the main lines, that many small lines appear in the emotional zone, and that the Head Line is inclined towards the Mount of the Moon.

As far as the suitability of hyper-intellectual people for analysis is concerned we must find out whether their " intellectualism " is superimposed upon an empty, primitive soul or whether their super-intellectual attitude has been adopted as a superficial screen against the outside world intended to camouflage certain powers —subconscious and inhibited forces, suppressed intuitive powers —which can be released and directed into proper channels by psycho-analytical treatment. A primitive mind precludes the possibility of development ; analysis, however, implies it, for it is the function of analysis to give inner stability, to help a person to find the centre of his being, and to develop his personality in the widest sense.

It would be wrong in the case of a neurotic illness to assume that only a certain part of an otherwise healthy mind is affected and can be treated separately : the person as a whole is neurotic. Thus if the psycho-therapist is able to liberate and render effective all the latent individual and collective unconscious forces it is most probable that the neurotic symptoms will disappear as well.

The necessary conditions for a successful analysis are patience, staying power, preparedness, sincerity and the will actively to support the analyst. It is of vital importance that the person being analysed should play fair, particularly in connection with dreams which may reveal weaknesses and unpleasant things, that he should not suppress important phases of such dreams or invent false ones just because he does not want to admit his failings to others and to the analyst : for dreams are the conscience of the soul. The inner acceptance of one's negative traits, which is of such vital importance and is achieved by the analysis, overthrows their dominance and shows the way to their elimination.

The argument that a person is perfectly aware of his difficulties and problems and that he knows all about them is by no means a plausible reason against analysis. On the contrary, an analysis, if properly conducted, transfers things from the intellectual sphere to the sphere of emotional and spiritual experience, and thus it is most important that those who are ripe for an analytic treatment but are reluctant to take the plunge should realise that analysis is not merely a method that can be grasped and fathomed by the intellect but an inner experience which affects the personality as a whole and leaves a deep and indelible mark on the person's innermost being.

Hands are also excellent indicators of talents and, therefore, particularly valuable factors in all matters of vocational guidance. As it is impossible adequately to deal with this branch of applied chirology in the limited space at my disposal, I can only discuss its general aspect. It happens very frequently that the hands of creative people, apart from the main lines and the Line of Intuition, show very few lines revealing personal experiences. The reason for this lies in the fact that for these people a personal experience is only a means to an end, that they consider it much less important than their creative work. I was, for instance, surprised to find no trace of serious difficulties in the hands of the late pianist and composer, Eugen d'Albert, who had been divorced several times. Only one partner-relationship was recorded in his hand, as a matter of fact the longest of his life. He asked me why this was so and I suggested that he must have transmuted his many matrimonial experiences into musical creations. He confirmed this and showed me his diary, which contained evidence that he had created a new work after every divorce.

Hands which reveal little of individual-personal experiences are particularly often those of architects and sculptors. It is characteristic of their experiences that they only seem to affect the personality peripherally because the centre of their being is completely occupied with their work. Mostly such experiences only concern them so far as they can form and mould other people according to their own ideas.

Artistic talent can only be adjudged if the person has a Mount of Apollo. In addition to the remarks about the fingers in Chapter Four, I should like to say that also the third finger, which, in such hands, is usually very long, straight, expressive and spatular, plays an important part.

The little finger of sculptors and architects is rectangular-shaped.

Productive or creative musicians, such as composers, conductors and instrumentalists, must have a rectangular or spatular, fourth finger as well as the Mount of Apollo and the Line of Intuition which leads to the third finger, a flexible thumb and a straight, expressive and spatular third finger. If the end of the fourth finger is conical the musical talent is mostly not productive and formative but indicates a good ear for music, that is to say, the person remembers and is capable of rendering what he has heard, or else that he is inspired and elated by music.

Should the Head Line in a rectangular-shaped hand with a rectangular-shaped second finger end with a fine curve below the fourth finger, the person is gifted for natural sciences.

Technical abilities of all kinds can be deduced from the flexibility and the nimble expression of the fingers, a usually very straight and long little finger and a good technical Line of Intuition.

Medical talent is revealed by an emphasised upper Mount of Venus and a good relationship between the second and third fingers. The Mounts of Mercury and Apollo are mostly combined.

Often the hands show talents which are not sufficiently great to warrant a professional training. Yet they are quite essential because when exercised they very often release idle and shackled powers of the mind and provide an outlet which in many cases of general disappointment and lack of self-confidence produces an invigorating and stimulating effect on the personality as a whole.

CHAPTER ELEVEN

CHIROLOGY APPLIED TO THE HANDS OF CHILDREN

Now let us return to the starting-point and final practical goal of this book : the hands of children. First of all, the chirological interpreter of children's hands must keep in mind that the outside of the hand is not sufficiently developed until the tenth or twelfth year to be considered in an analysis. This factor depends, of course, upon the general physical development of a child, for there are cases when the outer hand is sufficiently mature before the above age limit is reached. In most instances, however, only the inner hand can be submitted to a chirological analysis. The best procedure is to study the course and the intensity of the main lines and their relationship to each other, bearing in mind that any part or the entire length of the Line of Fate may develop at any time up to the twentieth year. Thus a weakly defined Line of Fate in the hand of a child does not mean that it may not become stronger and more definite in the course of the child's development. Also the size of the Mounts and their relationship to each other is fairly important in a child's hands.

I have already pointed out in Chapter Eight that children are either born with few or with very many lines. If the main lines appear, and apart from this there is little movement in the hand, that is to say there are not many subsidiary lines and lines of experience, this must be interpreted as a good sign, for there are possibilities of a harmonious development. Should, however, a very great number of subsidiary lines and lines of experience accompany the main lines, we must assume that on the whole these children will come up against difficulties in their development, for—as pointed out in a previous chapter—they are hereditarily burdened with the unsolved problems of their parents. In such cases the analysis of the right hand is of considerable importance for the correct interpretation of the child's personality, because the right hand changes very slightly in the course of time, and thus at an early stage permits of far-reaching conclusions about the psychological influence of the parents on the child. Apart from establishing the fact whether certain qualities have been inherited from the parents, which can be done by a comparison of the lines of both hands, the analysis of the right hand very often proves that the difficulties in the development of the child about which the parents are complaining are not so much connected with the child's dispositions but rather with their own conflicts and unsolved problems. In such cases the psycho-therapeutic treatment of the parents would yield better and quicker results than the analysis of the child. Unfortunately it is mostly very difficult to convince the parents of the necessity of such a course, as in the very cases in which they are the cause of the inner difficulties of their children they are completely unaware and unconscious of their own weaknesses. And yet it is impossible to help these children without the co-operation of their parents. It is most essential that parents should know themselves to be free of false, ego-inspired ambitions, of vanity and prejudice, those deadliest enemies of an unimpeded and harmonious development of the child.

The reason why so many charming children who seem to be so close to the Divine and to Nature do not develop into human beings with fine minds is mainly due to the fact that these promising children are being pampered and spoilt to such a degree that they are led to regard themselves as the centre of the Universe. This attitude fosters an urge for power and domination which inevitably leads to clashes with the other individuals of the environment. Weaker personalities who are unable to assert themselves as a consequence of this impact become subject to feelings of displeasure, uncertainty and embitterment which tend to disfigure their originally beautiful souls. With forceful characters who are strong enough to establish their domination over others, the lust for power becomes the centre and sole aim of their lives and thus corrupts and gradually destroys the original nobility of mind.

Difficulties of development are not always conditioned by difficulties of the parents, for often a slow and belated development is caused by the natural disposition of the child. Usually this is chirologically indicated in the hands of the child by a Line of Fate which either starts late or has its origin in the Life Line but otherwise is normal and well developed. In such cases it is imperative that the parents should be patient, for if they are worried and anxious they will only upset and inhibit the child and slow down the process of its development more than ever. In the many years of my studies of hands of children

I have come to the conclusion that particularly those children whose development is belated and slow are able to preserve their vital energies and to achieve greater and more valuable results later in life than children whose development is too rapid and thus comes to an earlier standstill. Precocious children may make amazing intellectual progress up to the twelfth or fourteenth year, but the lines of their hands furnish evidence that their premature development has been achieved by an unwarrantable expenditure of energy, thus endangering their nervous strength and their entire inner life. The chirological sign of precocity is a Head Line which is already forked below the middle finger, and it is essential that the education of children who have this Head Line should be carefully supervised.

On the whole, experience has taught us that children whose hands contain a multitude of lines should be trained in handicraft; also gardening is often beneficial. On these children, who are inclined to be capricious, to lack concentration, to fritter away their energies, manual work will have a favourable effect because they will be producing something definite with their own hands and by their own efforts. Moreover, the continuity of their activity has a steadying influence.

It is of paramount importance that the mind of a child should grow, develop and mould itself slowly and gradually, daily assimilating some of the wonders of the surrounding world, and that it should not be introduced at too early a date to the world of rational knowledge.

It cannot be repeated too often and too emphatically that parents should free themselves of all personal ambitions, of vanity, of over-compensating wishes for their child, that they should try to discover and pay heed to the inner rhythm of its individual life, that they should be patient and not overactive. How often are valuable talents and dispositions of a child ignored or even suppressed because of the fear of its parents, who are afraid that their child is too playful and dreamy and therefore will never achieve anything in this world. Exaggerated activity on the part of the parents tends to frighten and discourage the child, to make it uncertain of itself, thus producing nervous disturbances and considerable difficulties of development and a gradual diminishing of its original gifts and dispositions. An analysis of the hands with its possibilities of comparison of individual and hereditary dispositions and a subsequent frank and candid talk with the parents, and guidance can do a great deal of good.

In this the psycho-chirologist should see his finest and most noble task, which is reward in itself: to be a helper in the shaping of growing life.

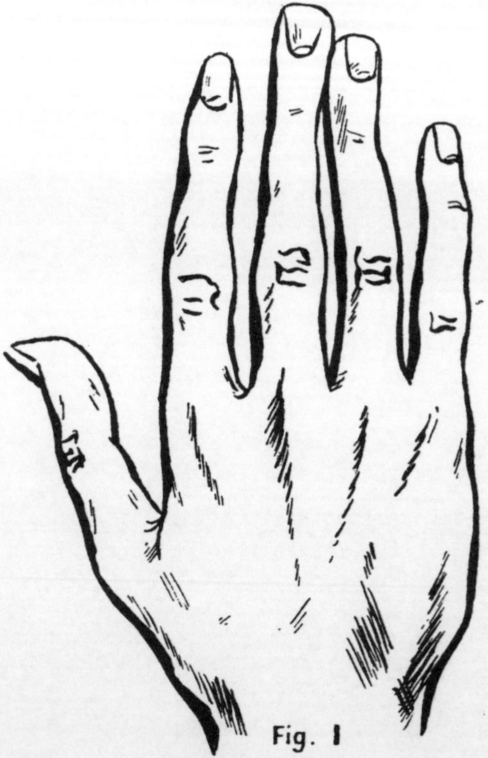

Fig. 1

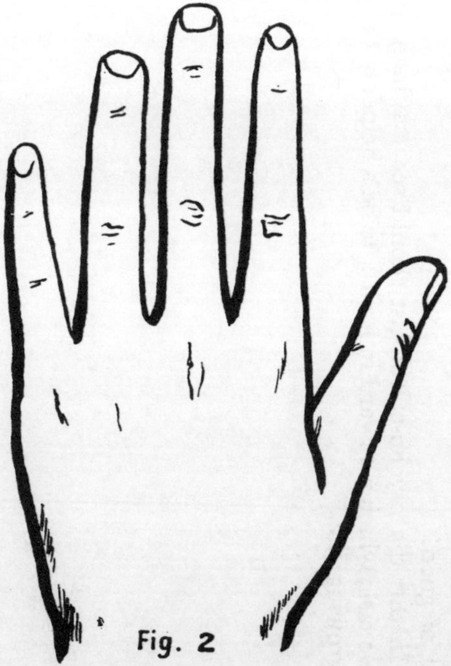

Fig. 2

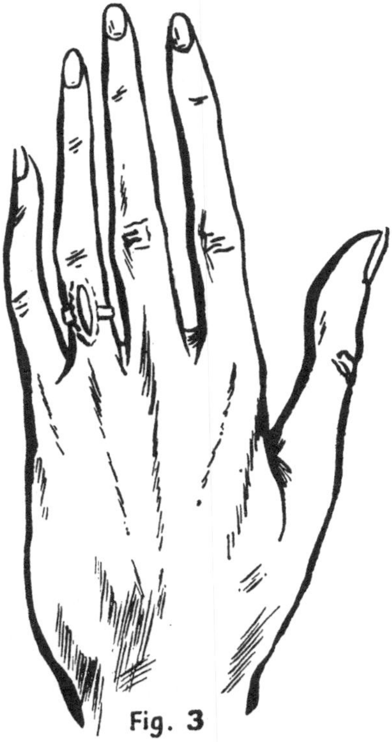
Fig. 3

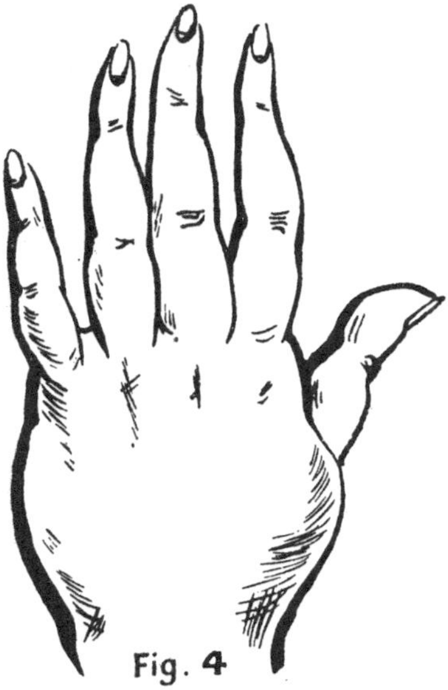
Fig. 4

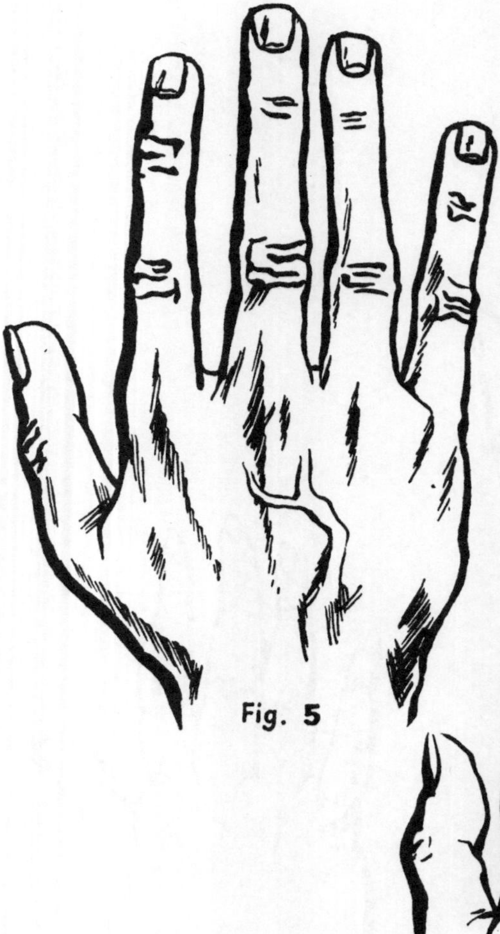

Fig. 5

Fig. 6

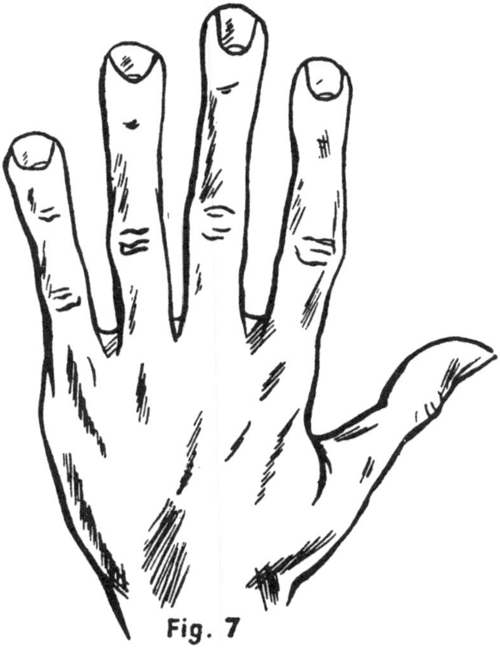
Fig. 7

Fig. 8

Fig. 19

Fig. 20

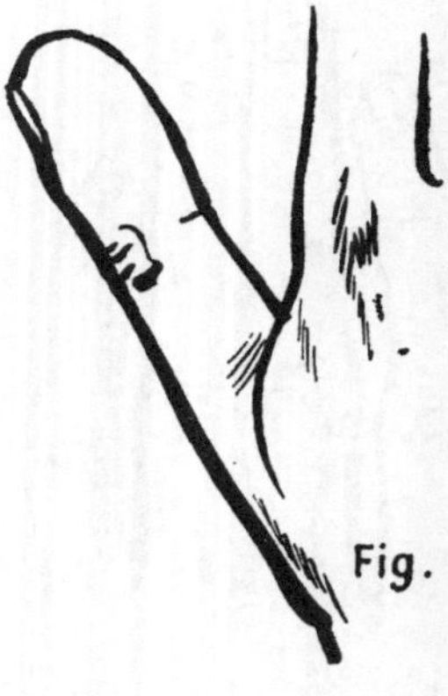

Fig. 21

Fig. 22

Fig. 23

Fig. 24

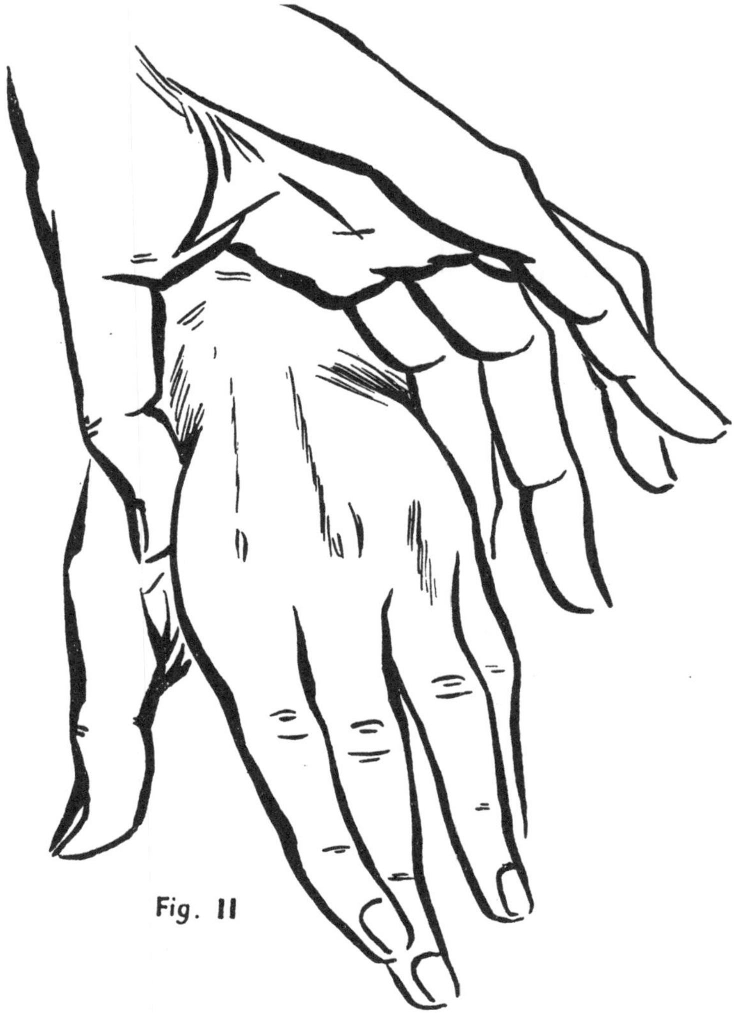

Fig. II

Fig. 12

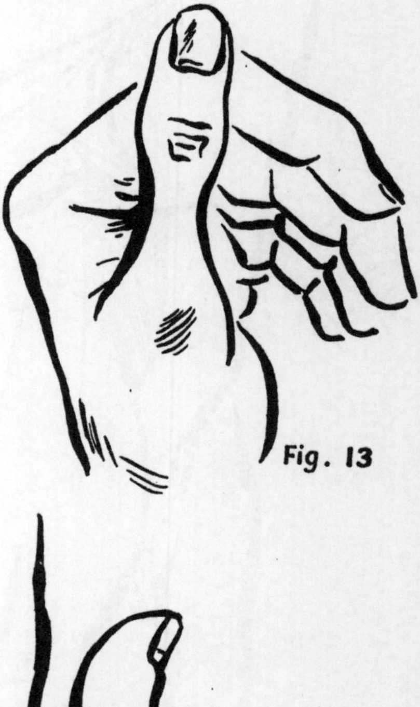

Fig. 13

Fig. 14

Fig. 15

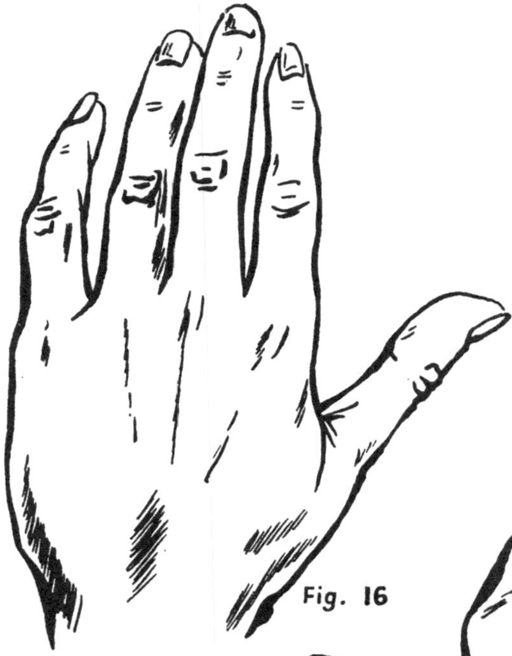

Fig. 16

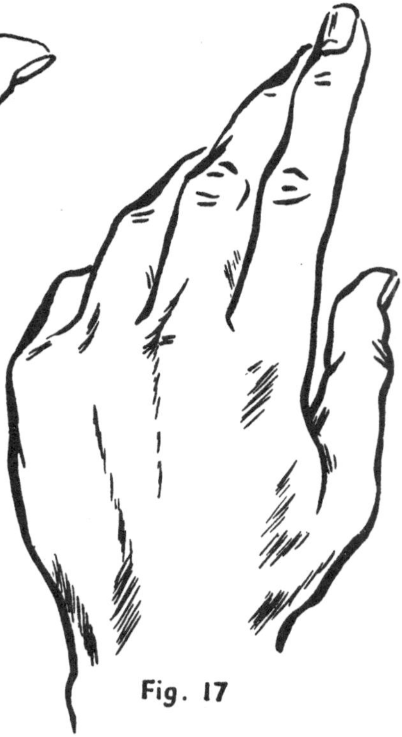

Fig. 17

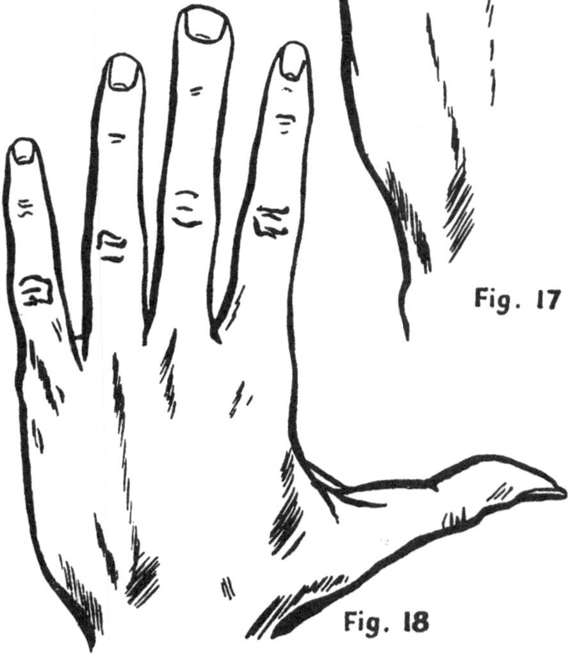

Fig. 18

Fig. 19

Fig. 20

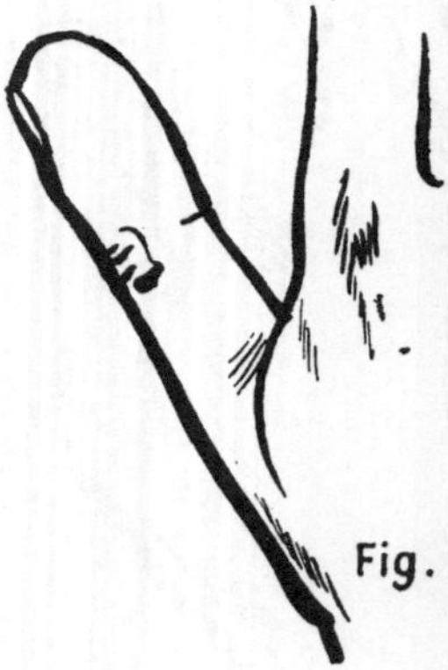

Fig. 21

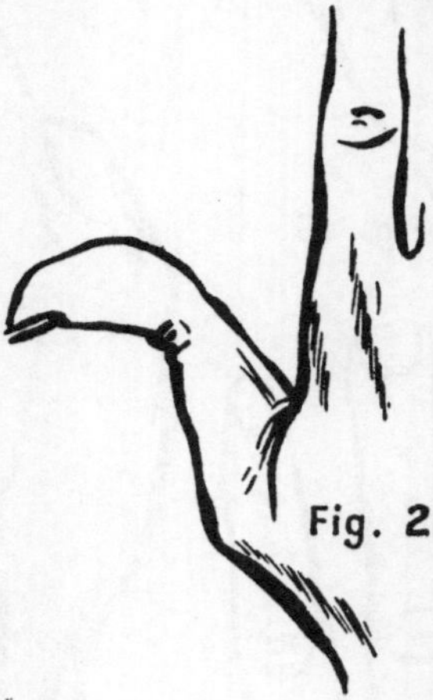

Fig. 22

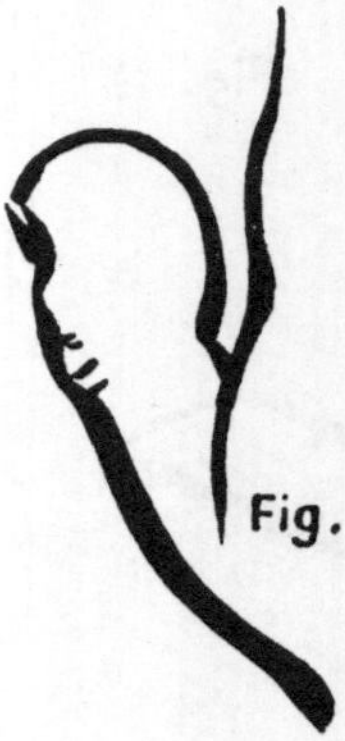

Fig. 23

Fig. 24

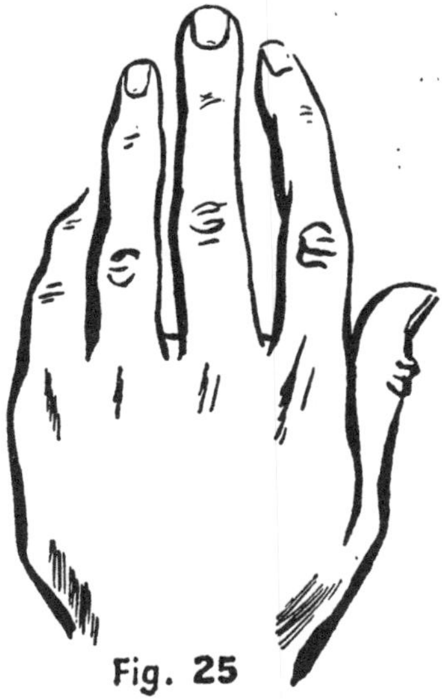

Fig. 25

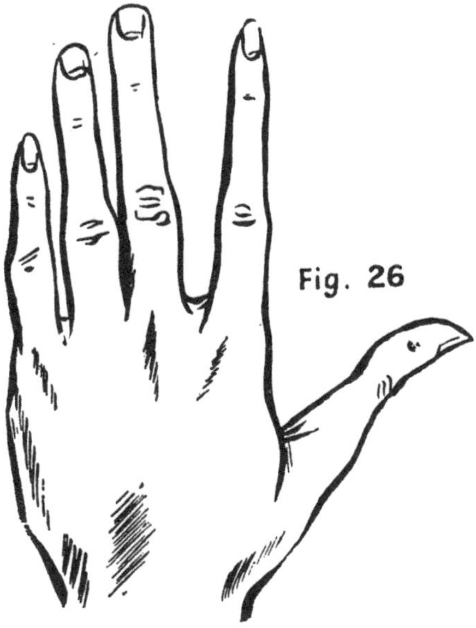

Fig. 26

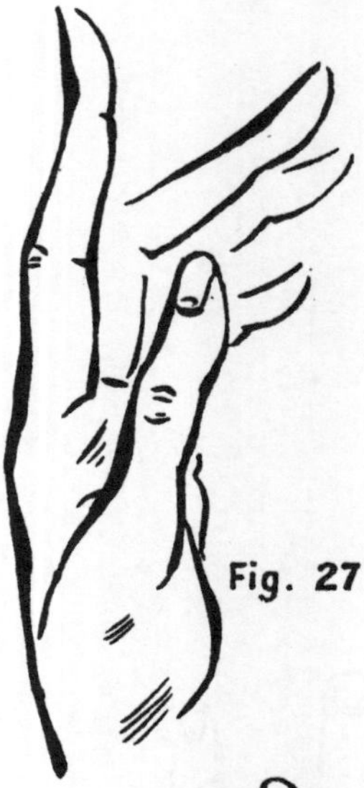

Fig. 27

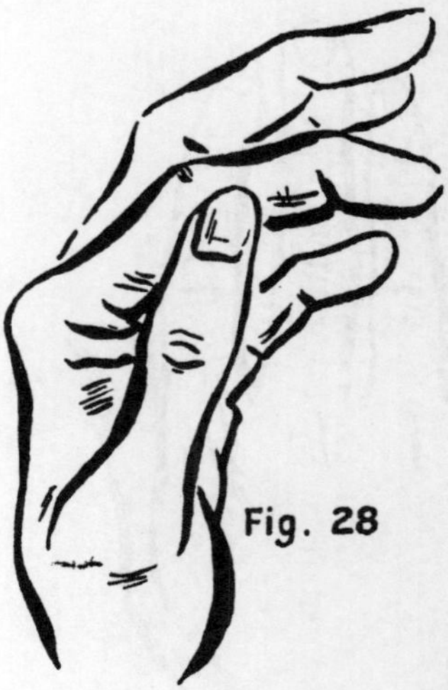

Fig. 28

Fig. 29

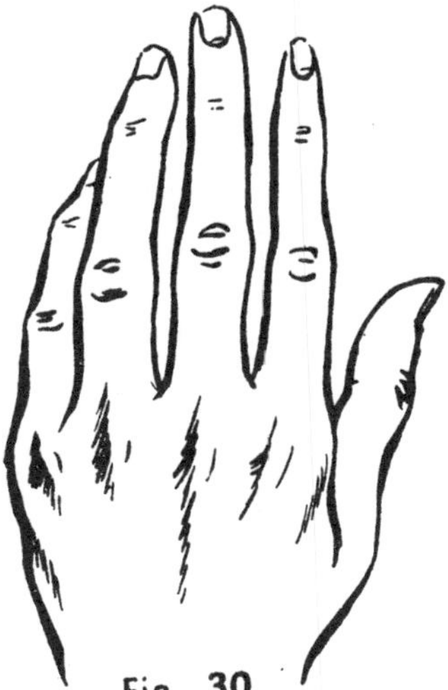

Fig. 30

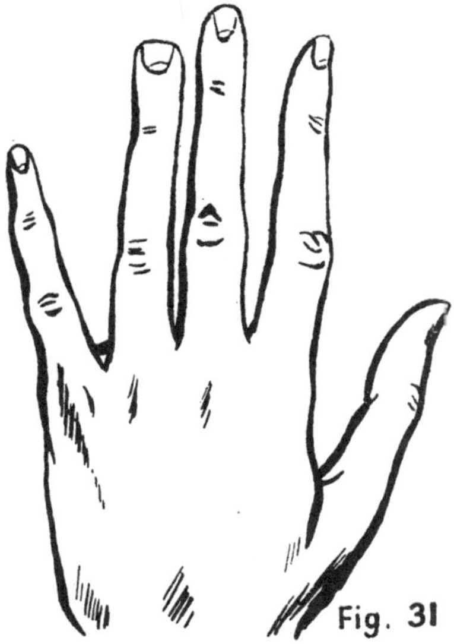

Fig. 31

Fig. 32

Fig. 33

Fig. 34

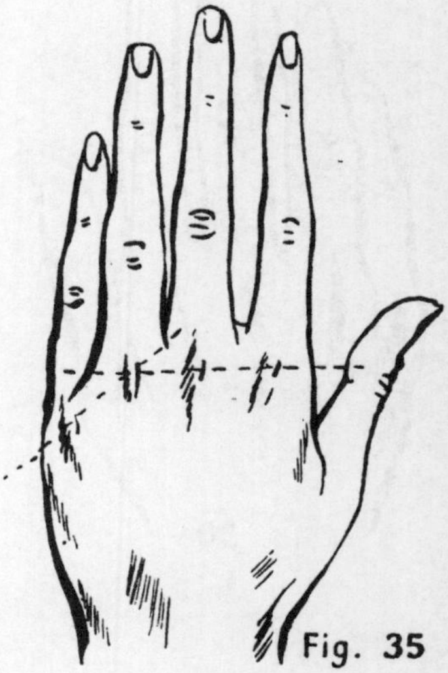

Fig. 35

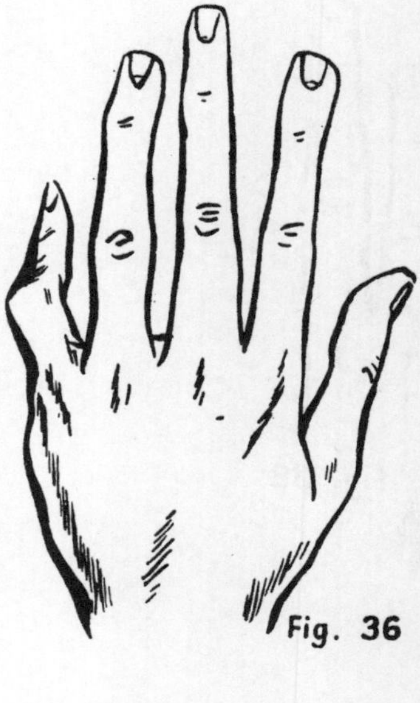

Fig. 36

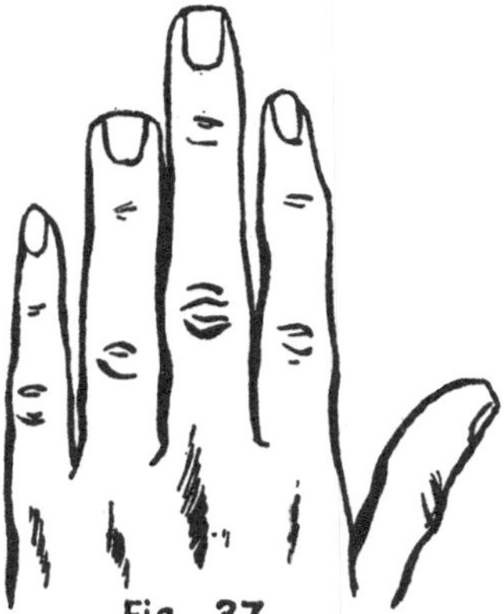

Fig. 37

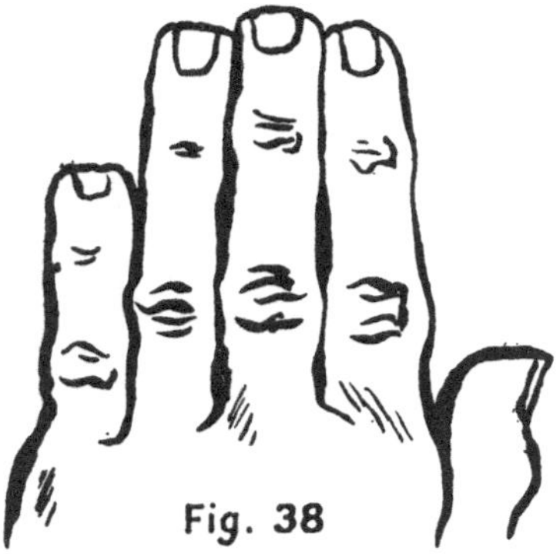

Fig. 38

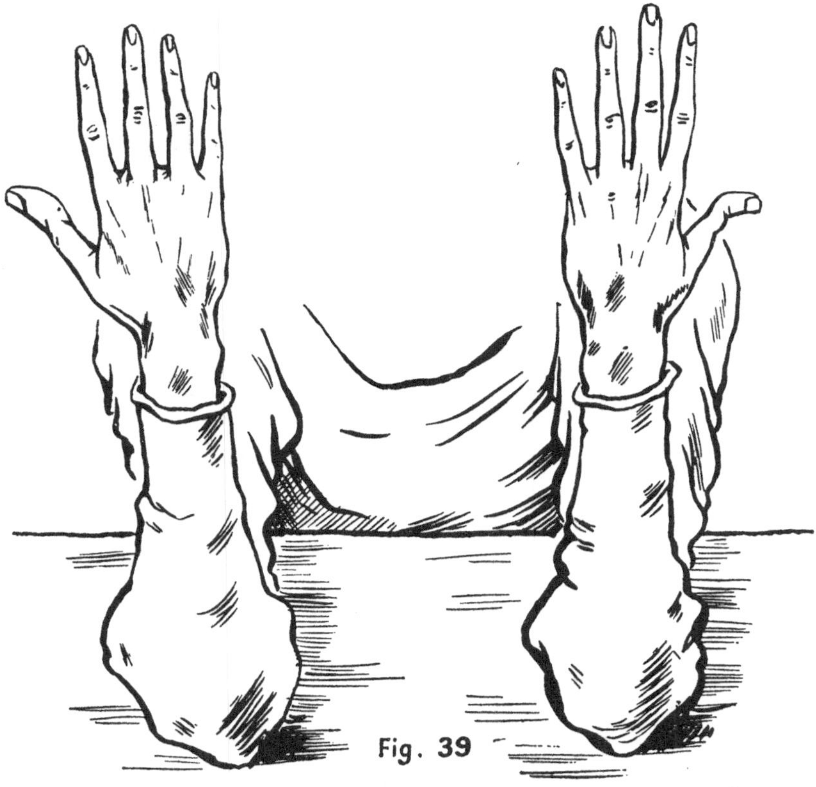

Fig. 39

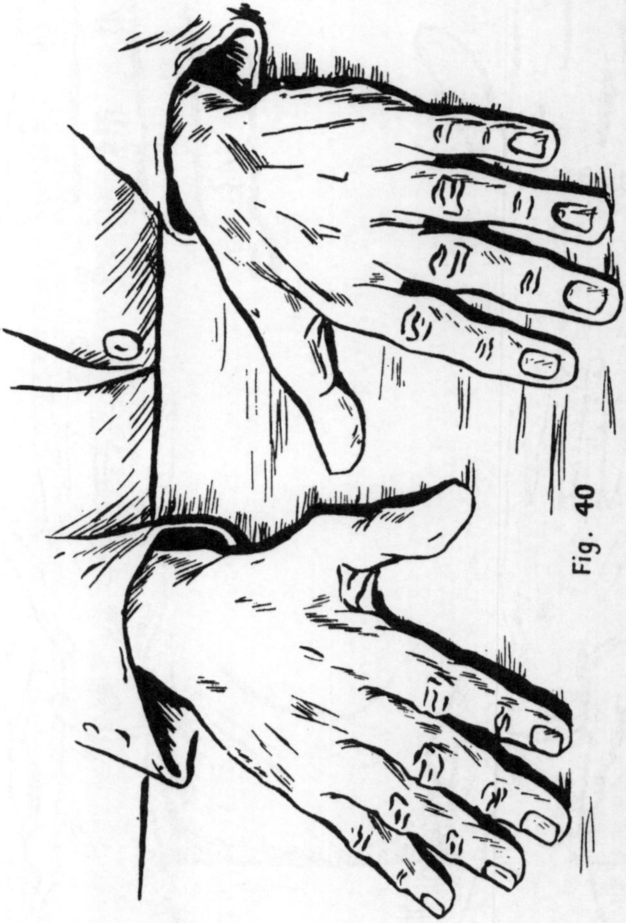

Fig. 40

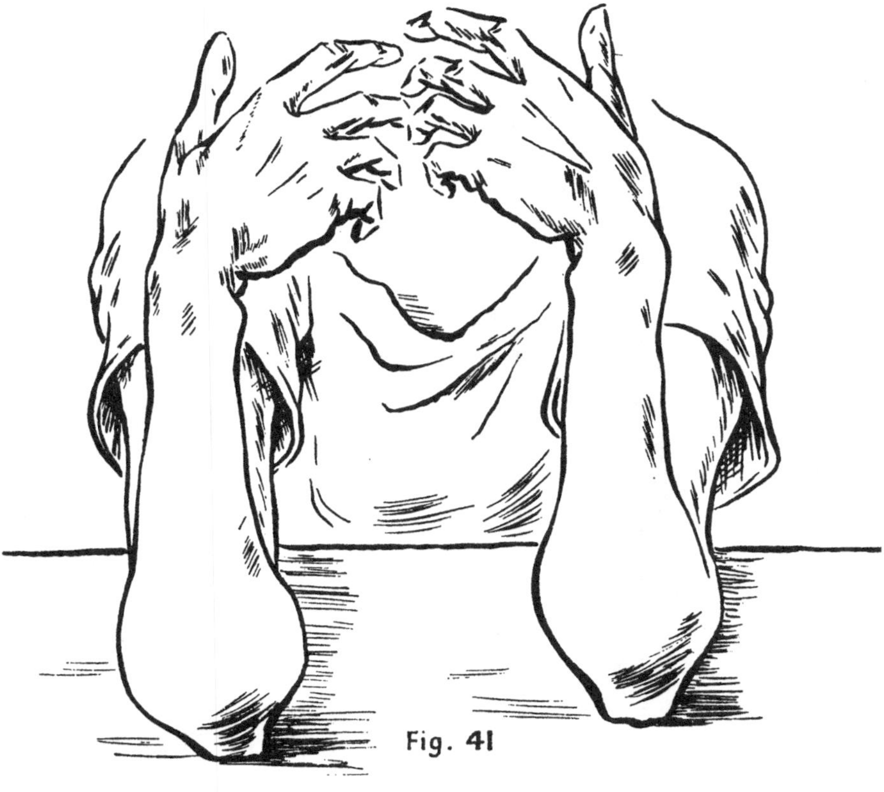

Fig. 41

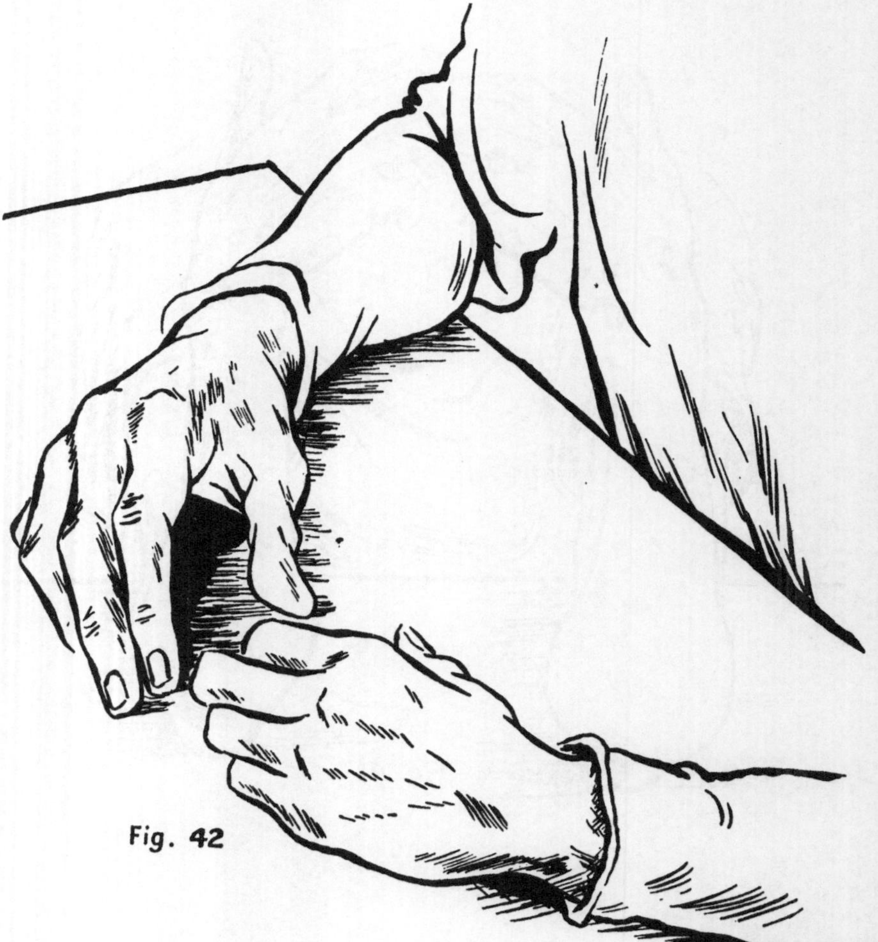

Fig. 42

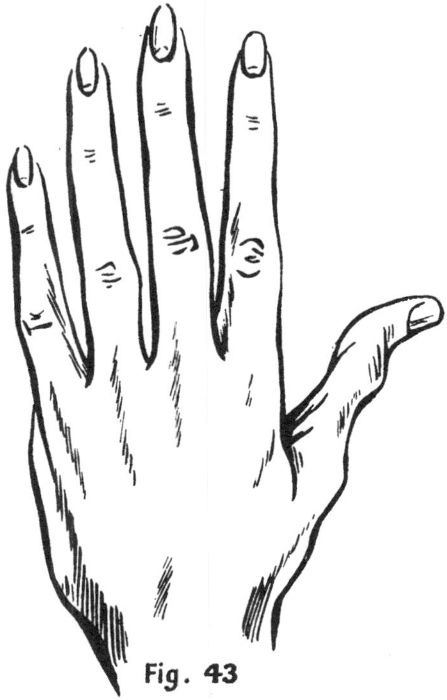
Fig. 43

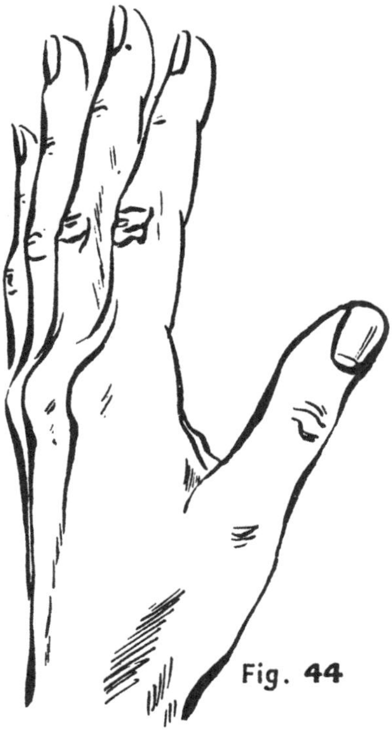
Fig. 44

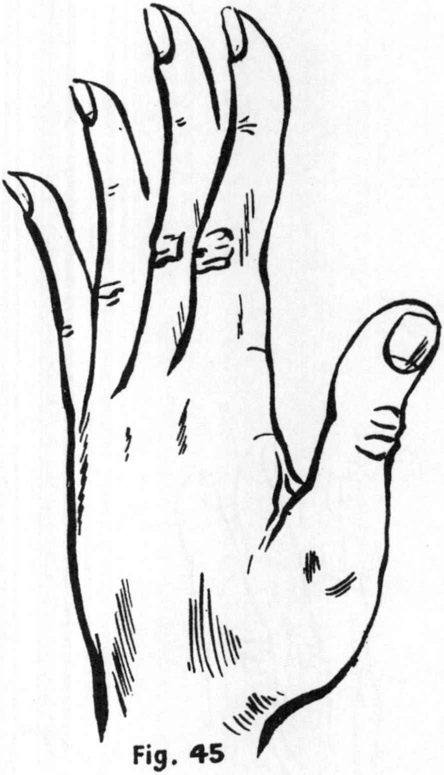

Fig. 45

Fig. 46

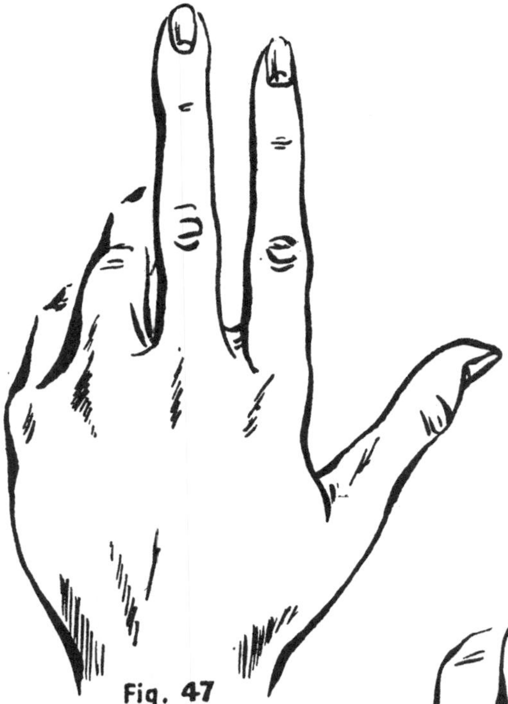

Fig. 47

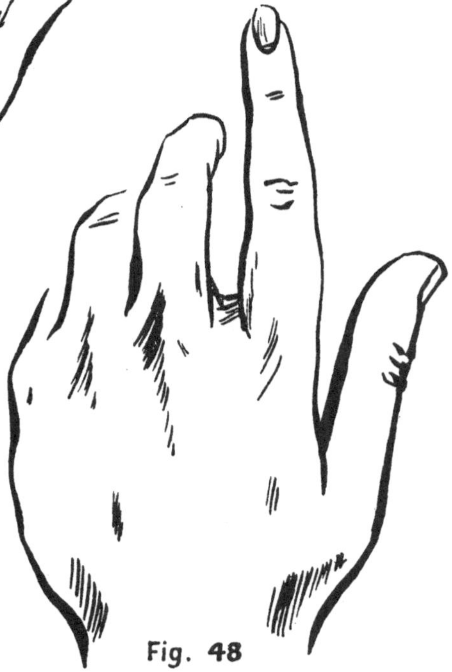

Fig. 48

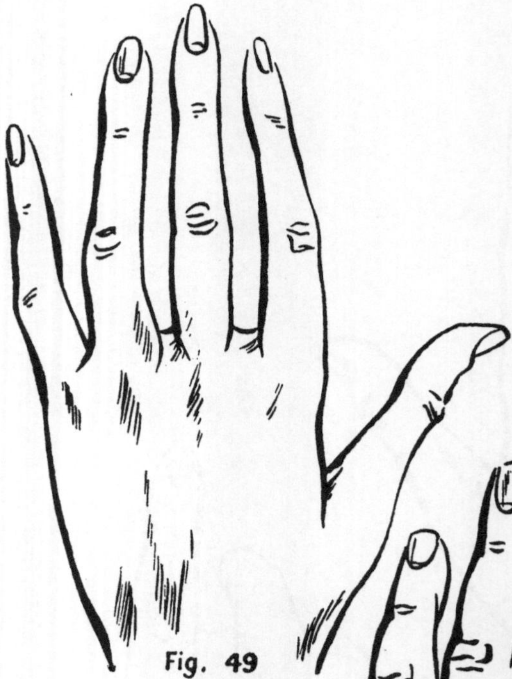

Fig. 49

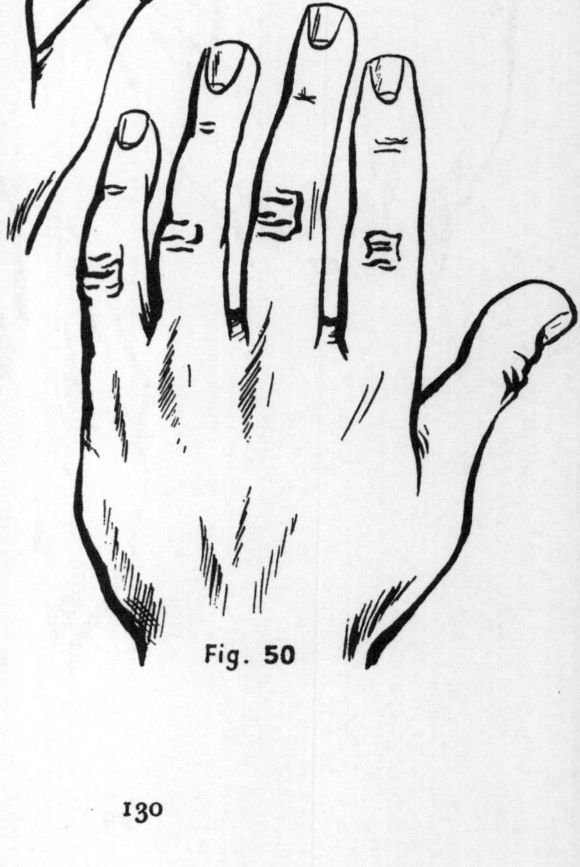

Fig. 50

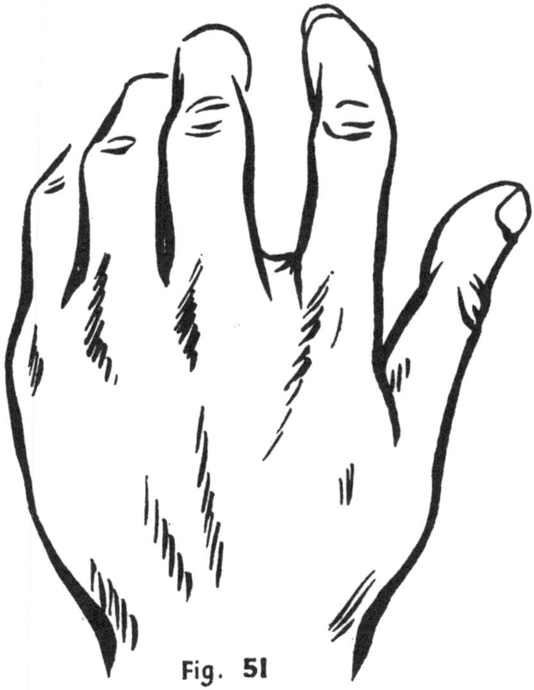

Fig. 51

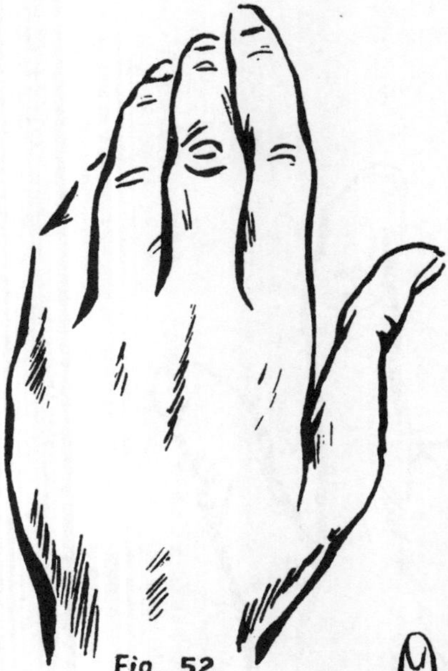

Fig. 52

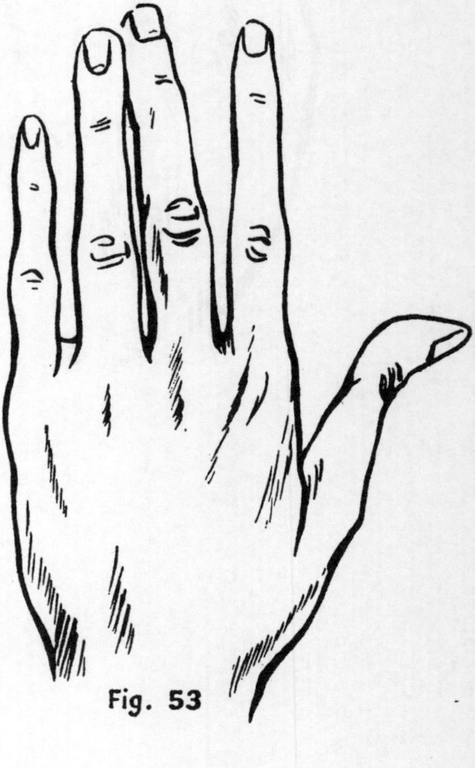

Fig. 53

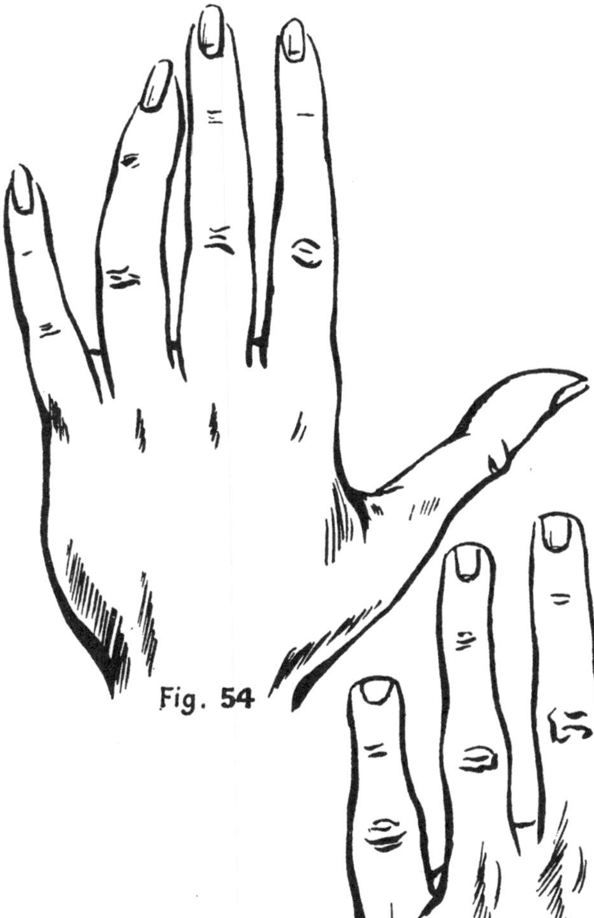

Fig. 54

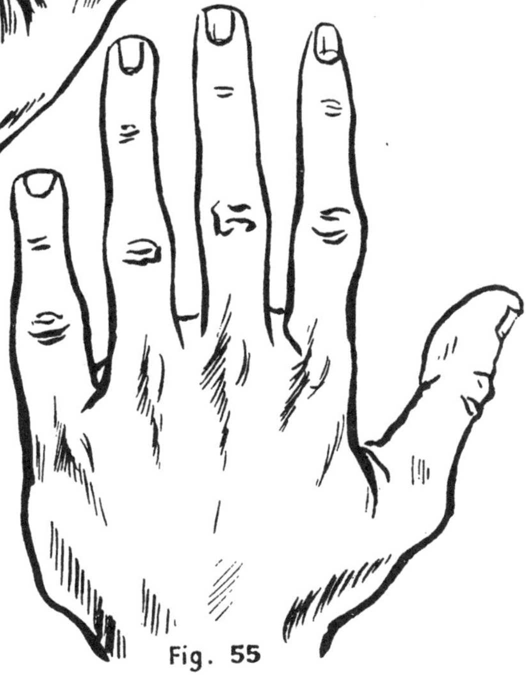

Fig. 55

Fig. 56

Fig. 57

Fig. 58

Fig. 59

Fig. 60

Fig. 61

Fig. 62

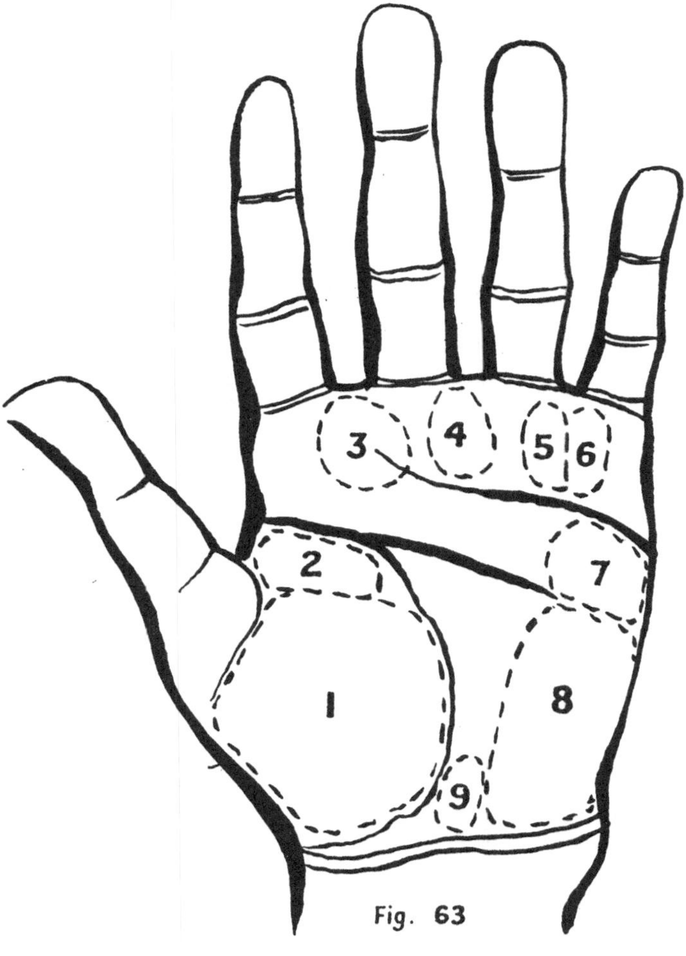

Fig. 63

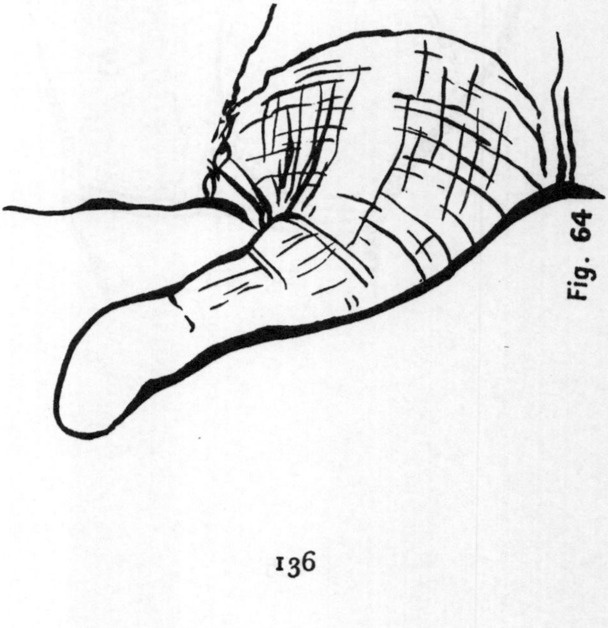

Fig. 64

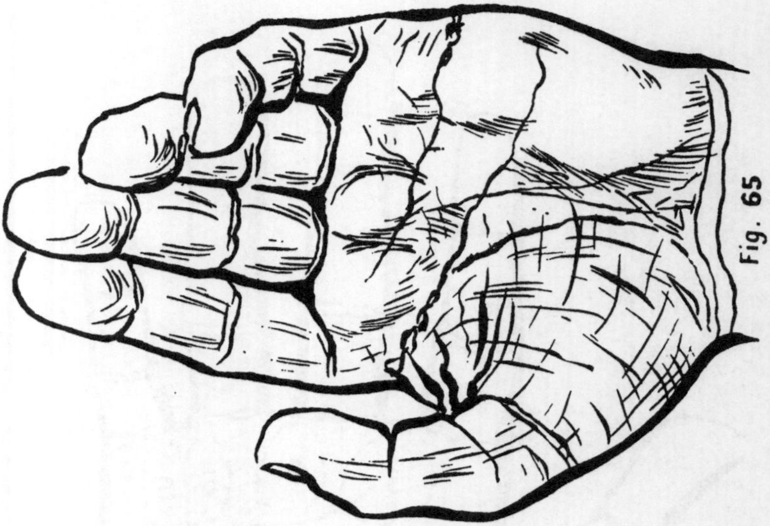

Fig. 65

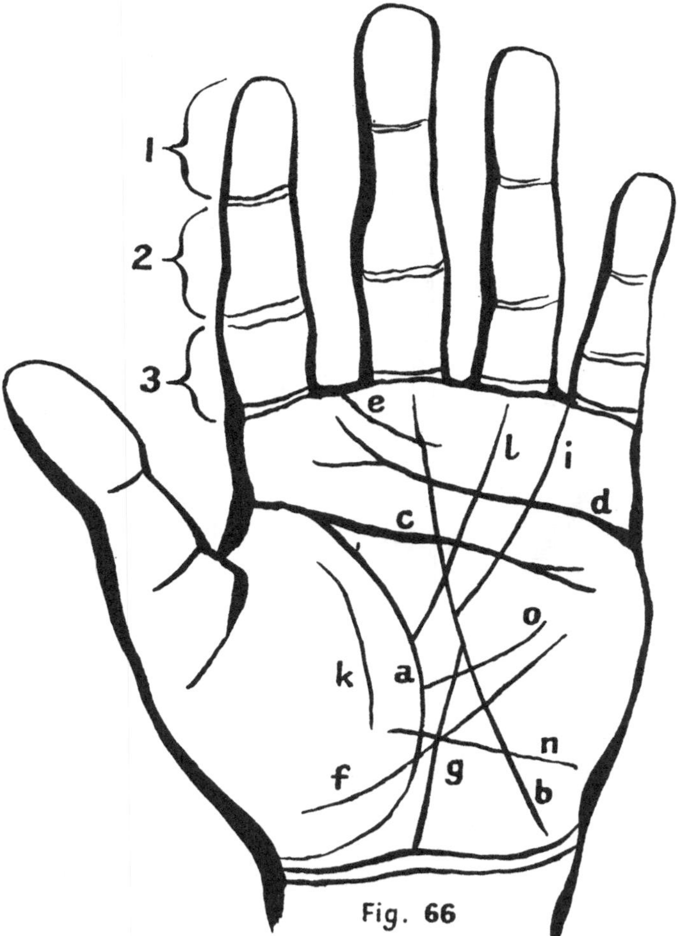

Fig. 66

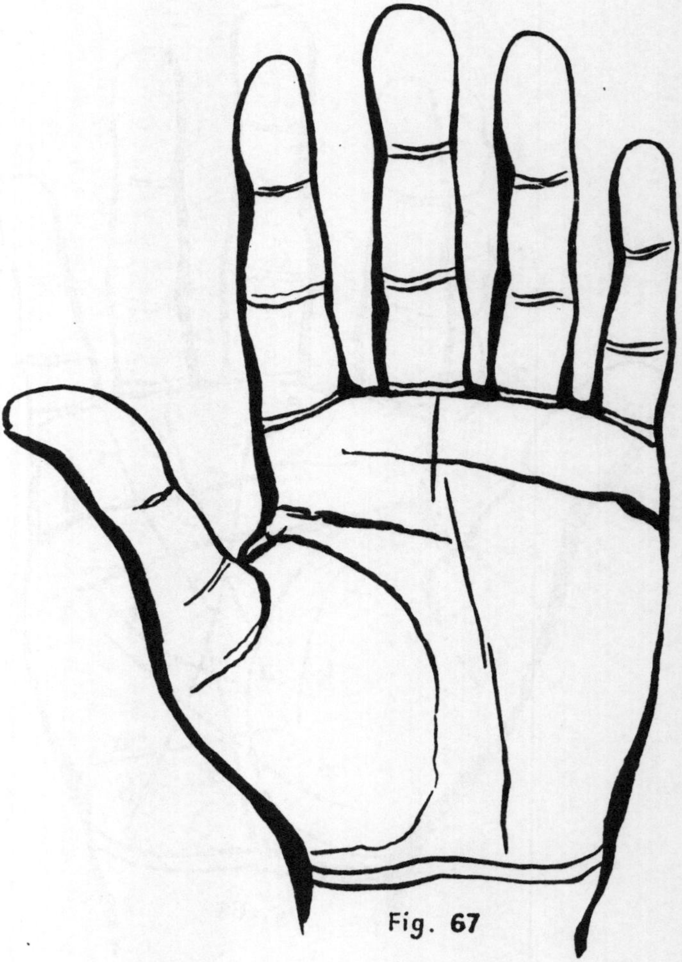

Fig. 67

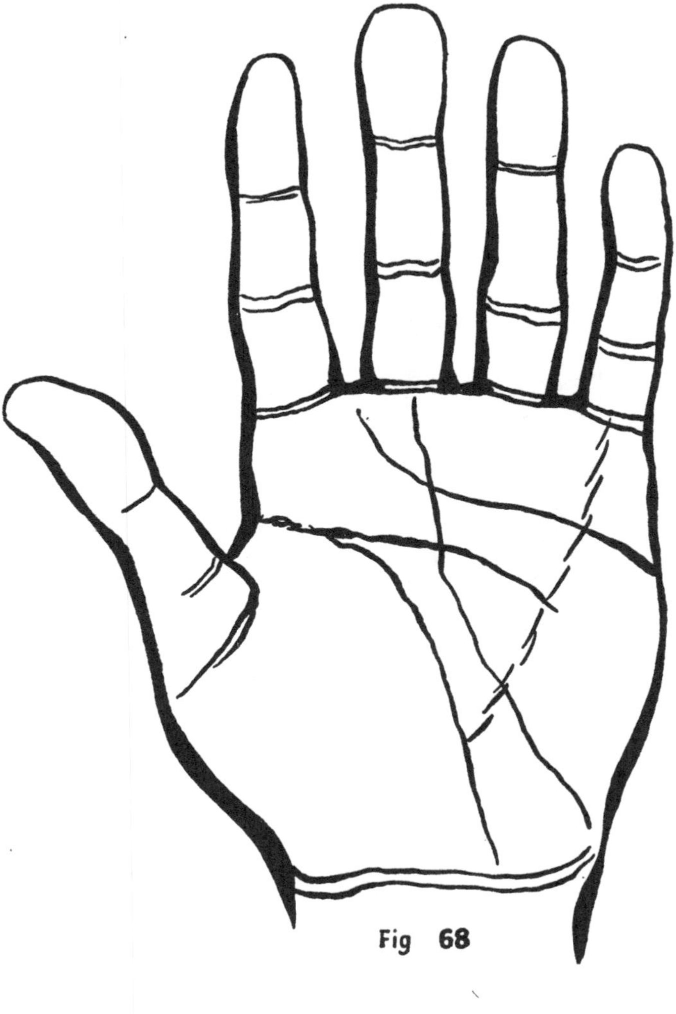

Fig 68

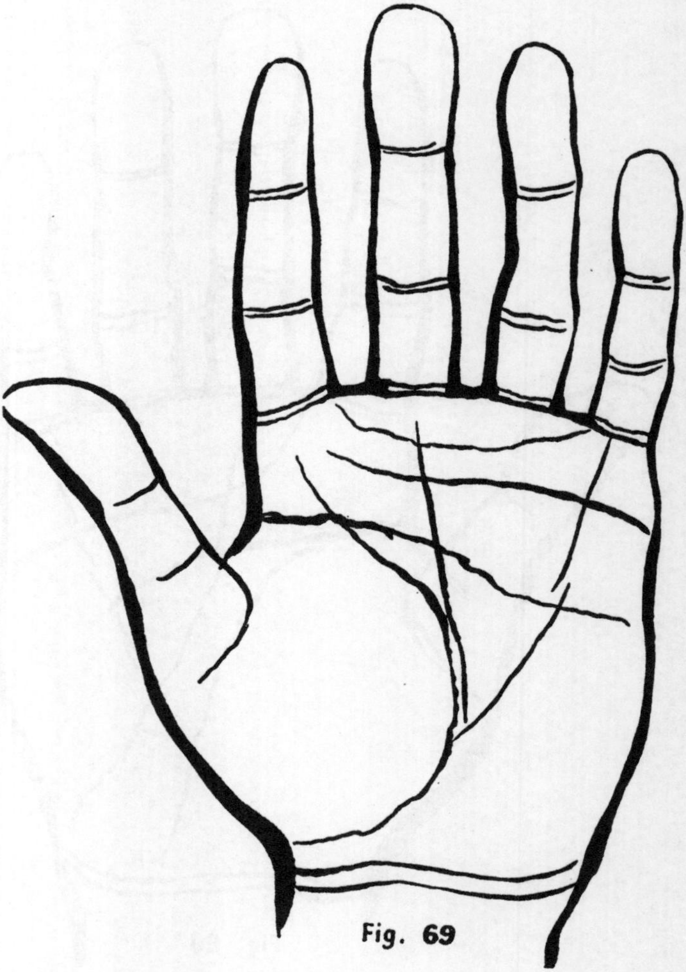

Fig. 69

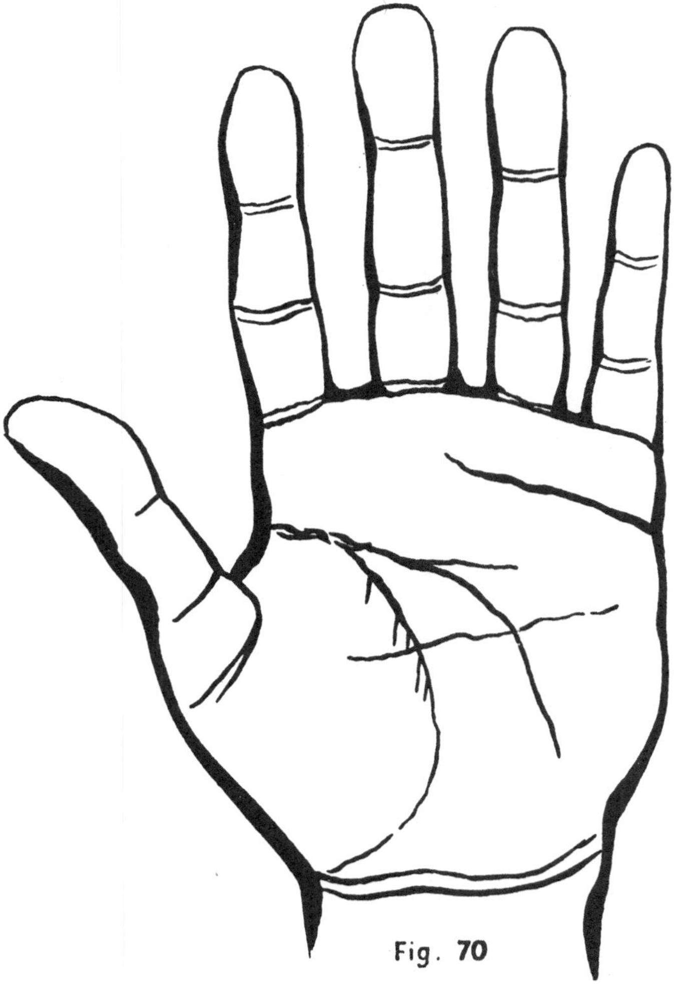

Fig. 70

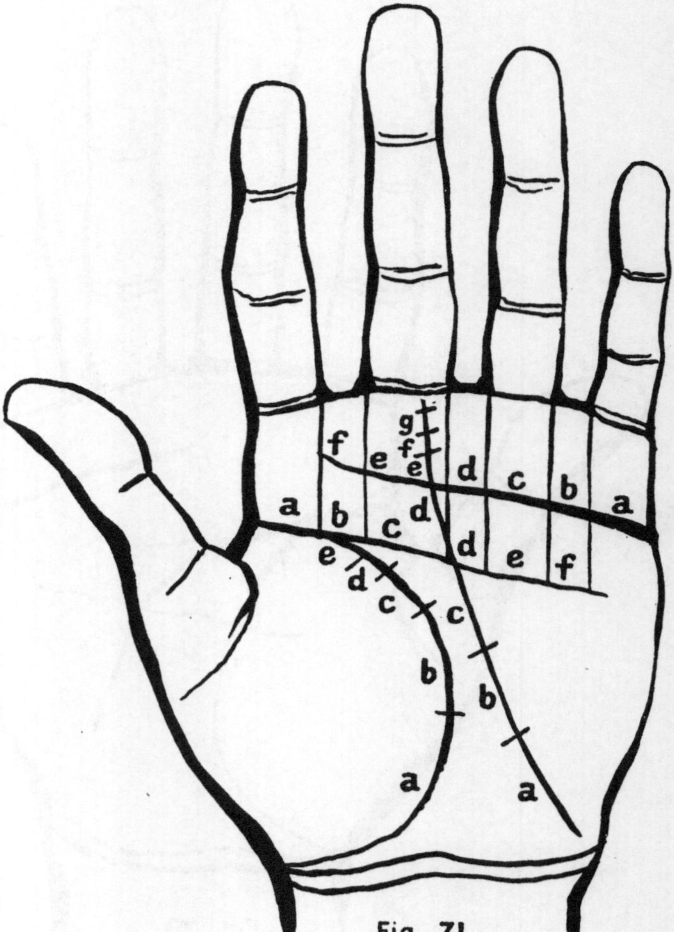

Fig. 71

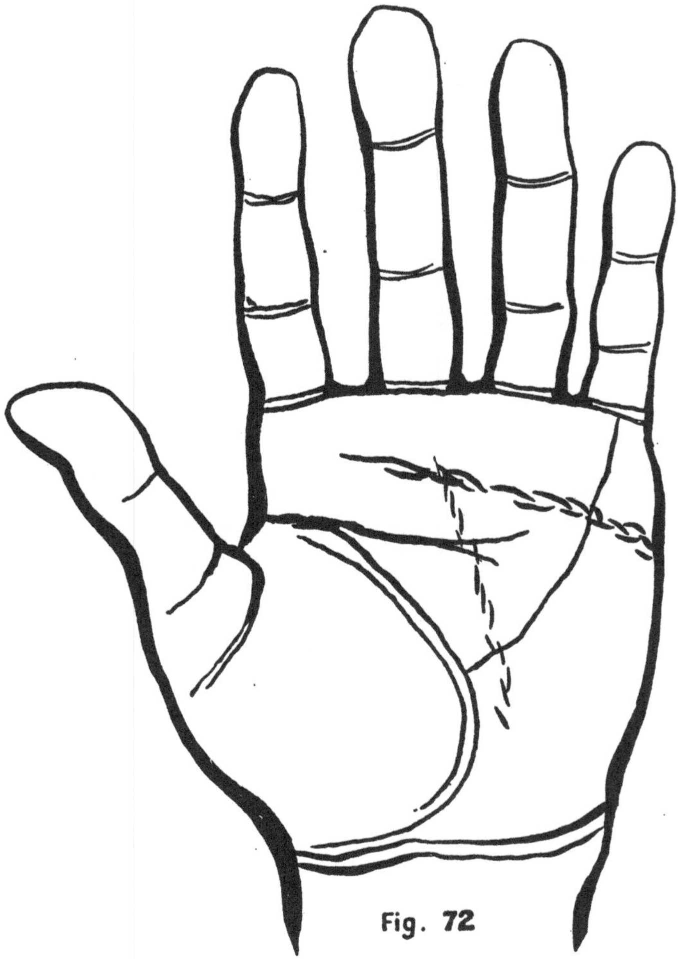

Fig. 72

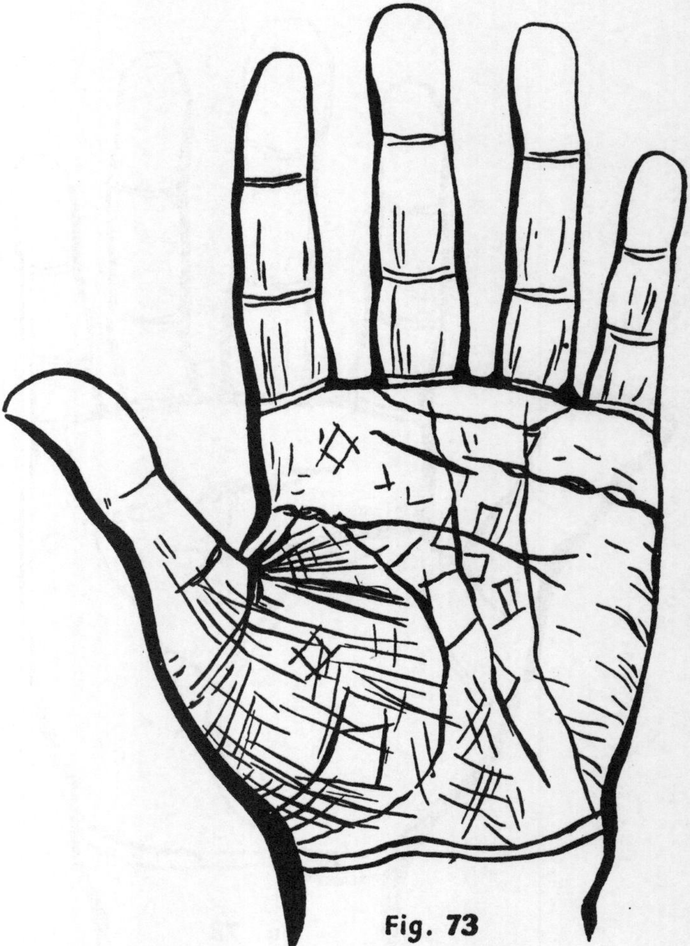

Fig. 73

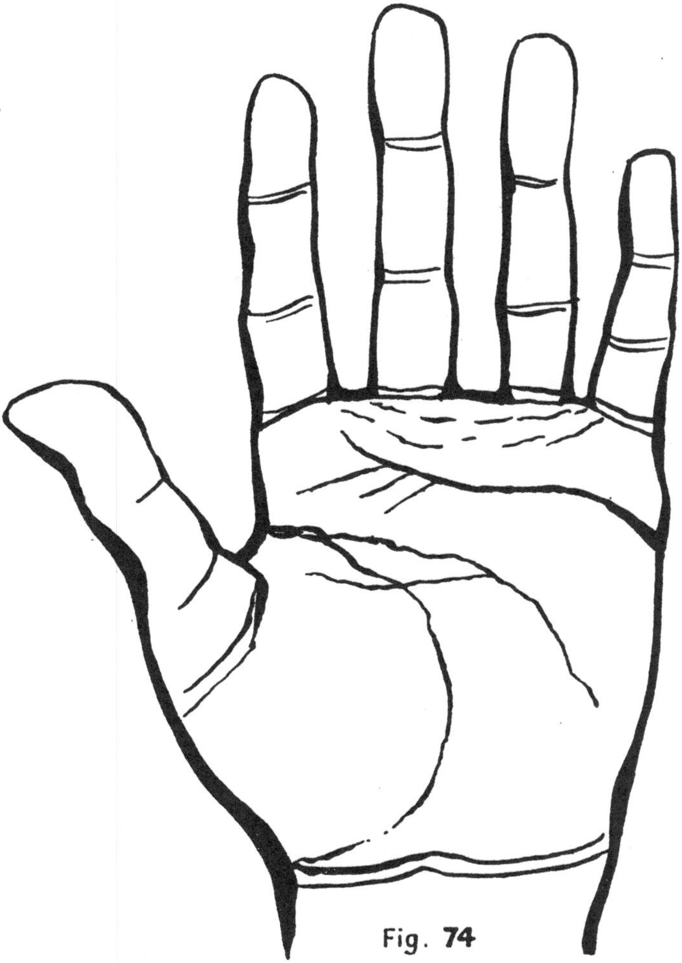

Fig. 74

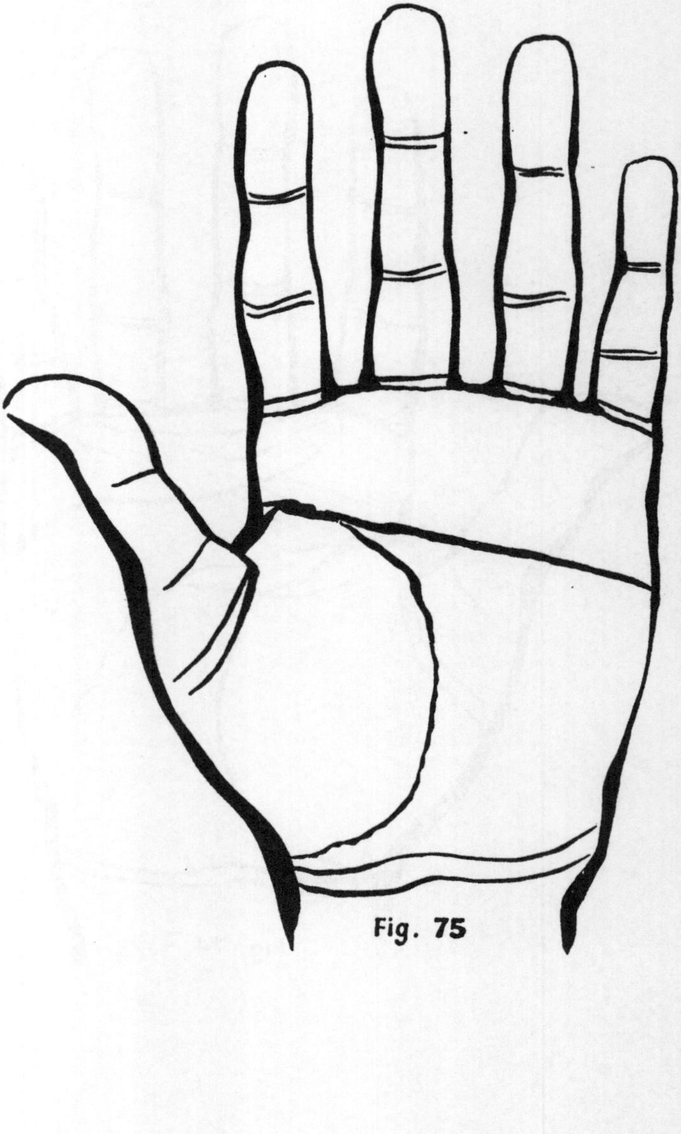

Fig. 75

Fig. 76

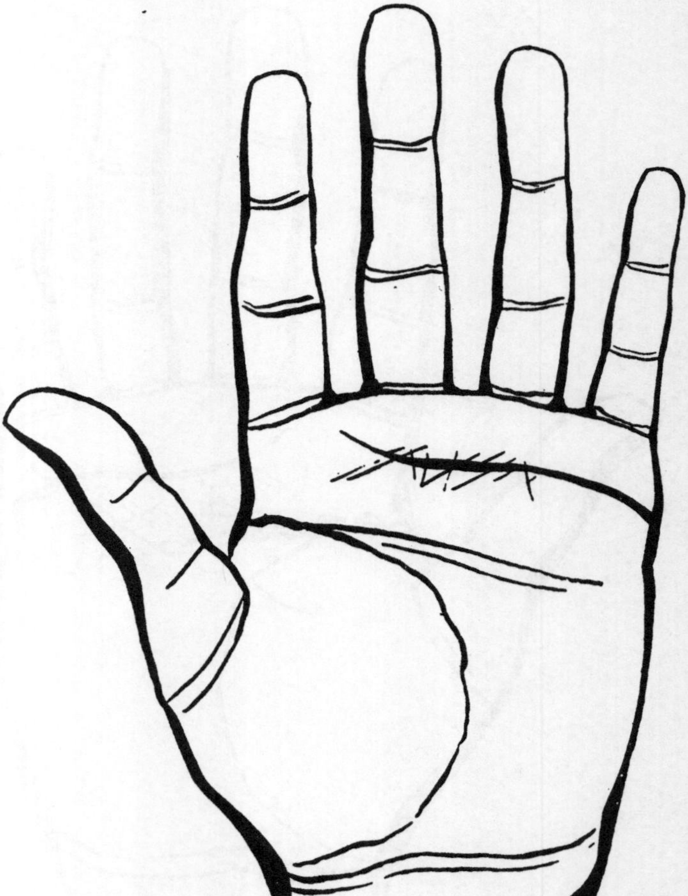

Fig. 77

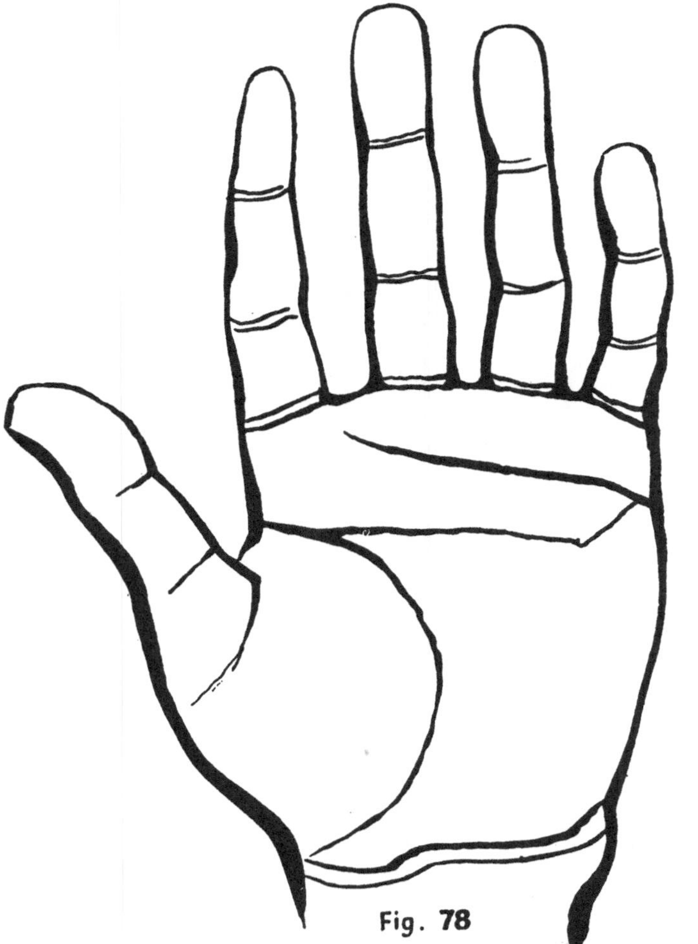

Fig. 78

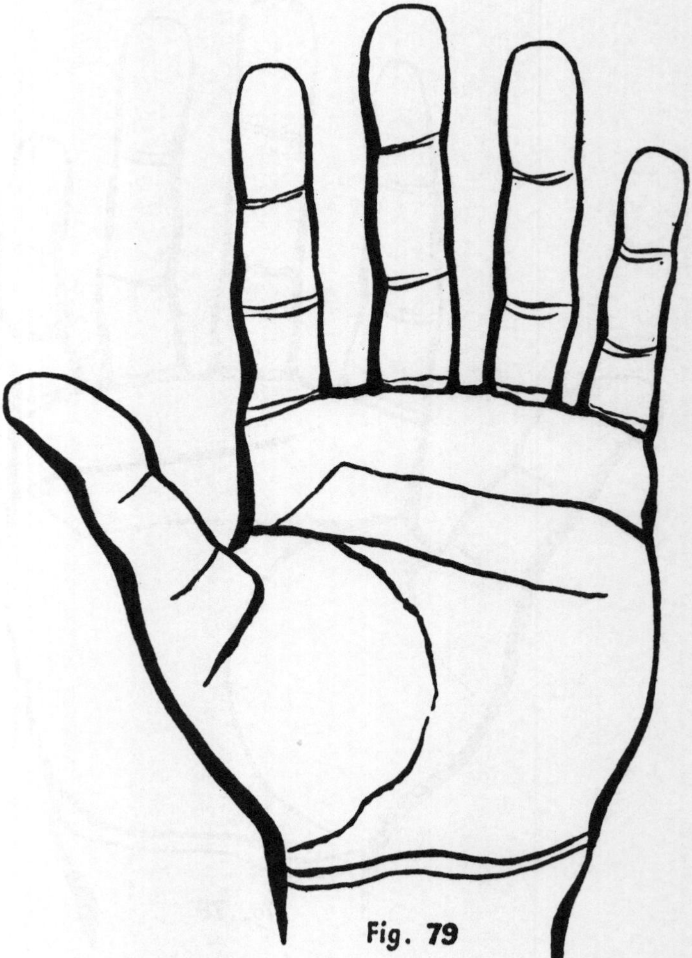

Fig. 79

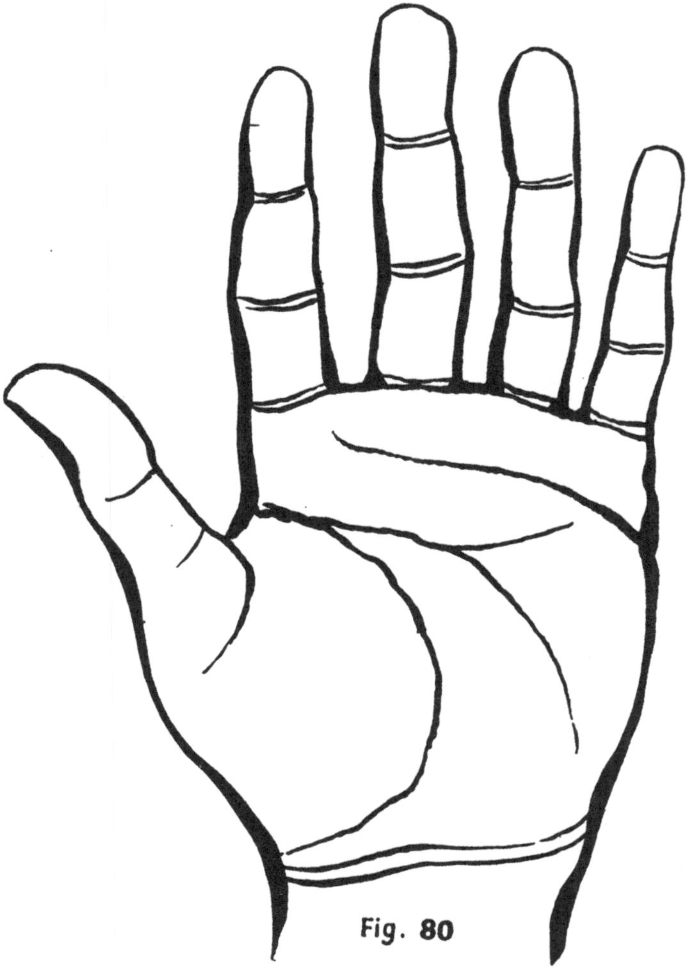

Fig. 80

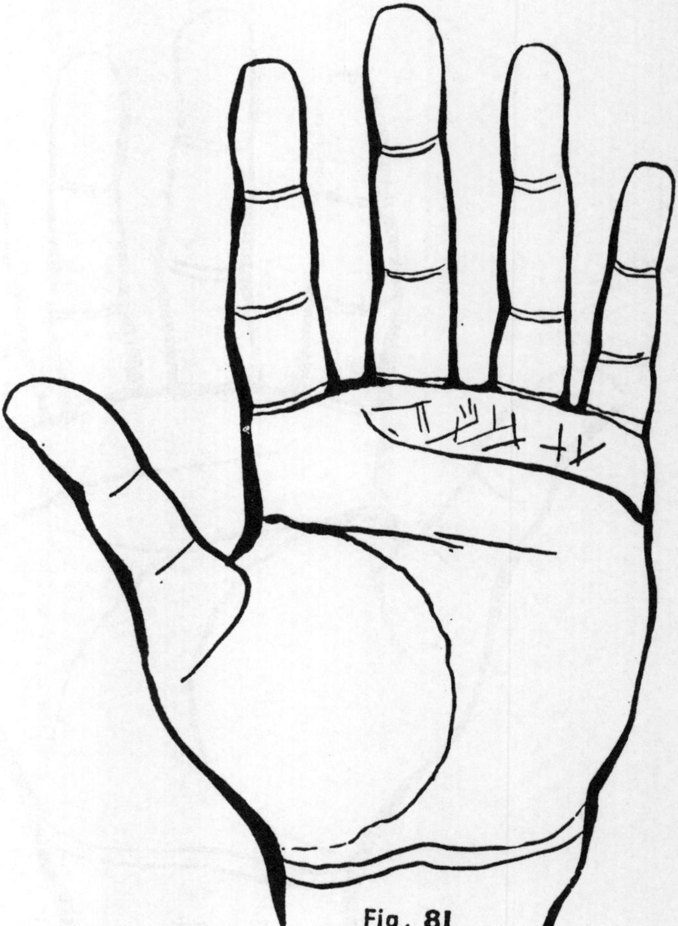

Fig. 81

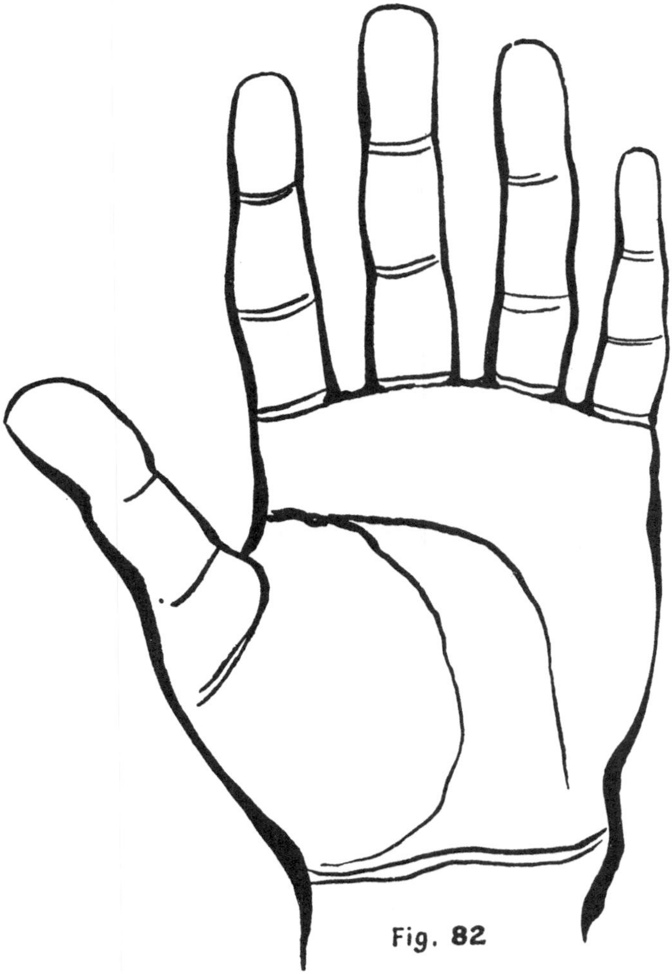

Fig. 82

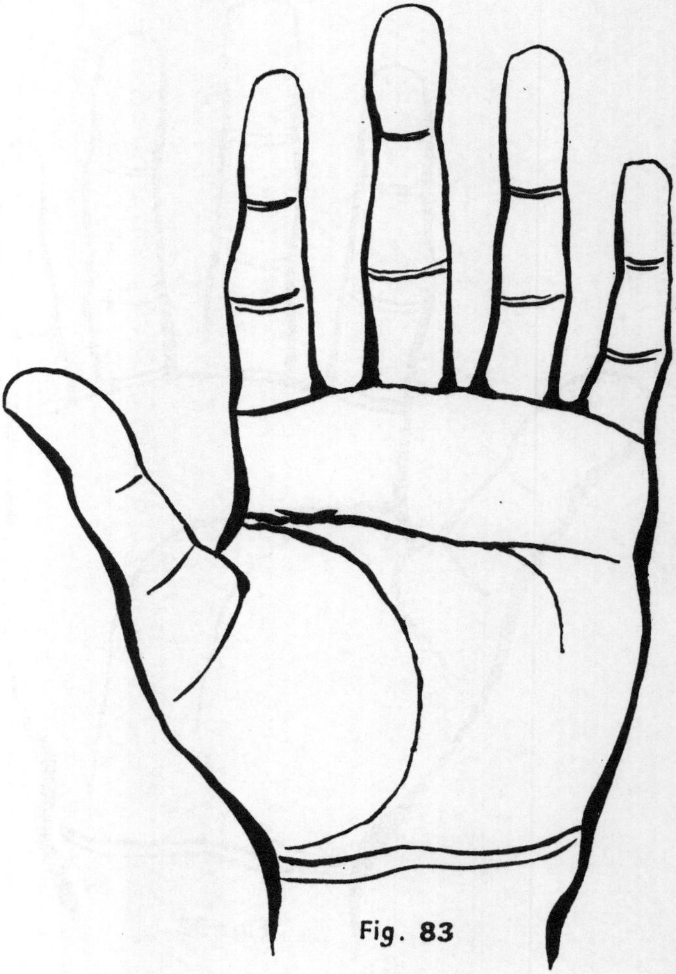

Fig. 83

Fig. 84

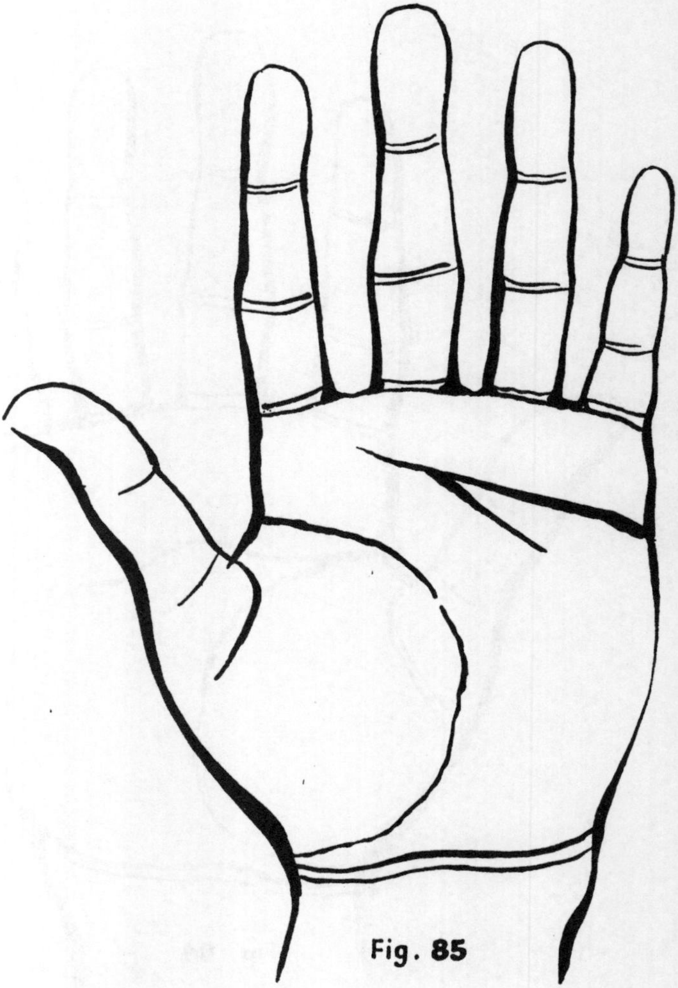

Fig. 85

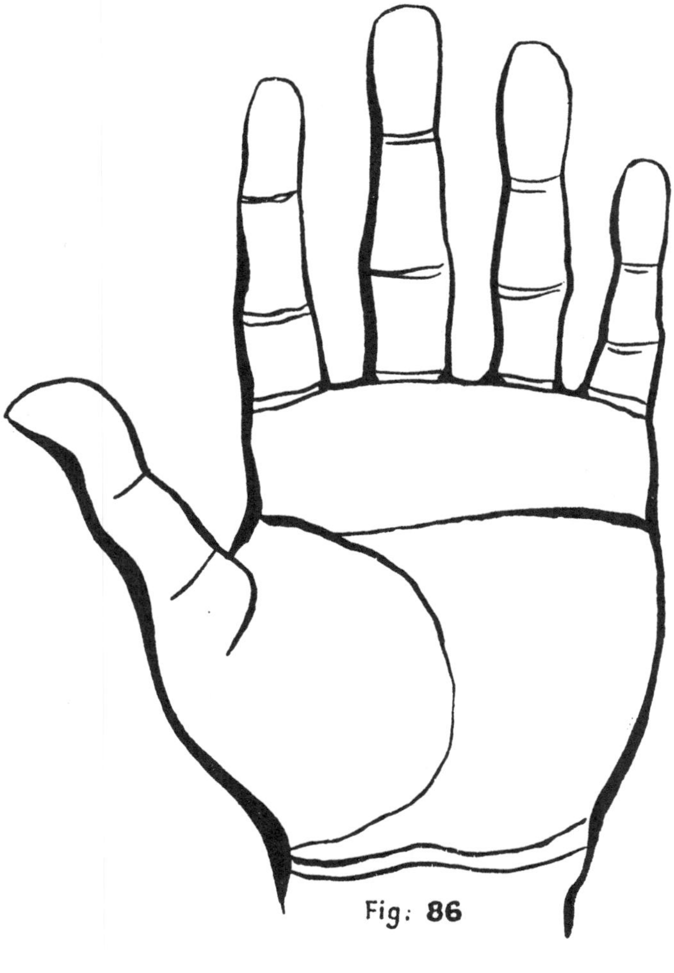

Fig: 86

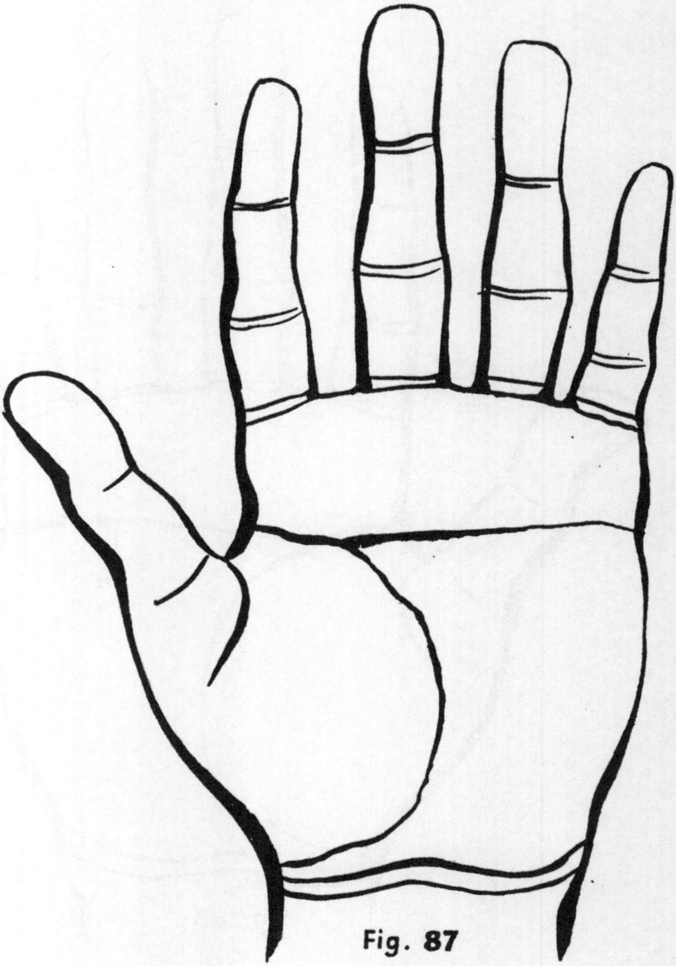

Fig. 87

APPENDIX ONE

SPECIMEN ANALYSES OF HANDS OF CHILDREN

I. The Hands of a Boy of Twelve

In both hands only two main lines appear, the Life Line and the Heart Line. The Head Line and the Line of Fate are missing. According to its form the hand belongs to the Primitive Type, and its expression is that of complete lifelessness.

The absence of two important main lines shows that the boy is mentally deficient, an imbecile who is incapable of development. (See plate I.)

II. The Hands of a Boy of Nine

In both hands the Line of Fate is missing, though in contrast to the previous case there is a Head Line. It is very short in either hand. The aspect of the hand is on the whole livelier, particularly the Mount of Venus, and there is also a Line of Sensitiveness (*e*). The boy is mentally backward and feeble-minded but, within limits, capable of development. He is very sensitive, and in contrast to the idiot boy of twelve he feels that he is not as bright as other children. Apart from line *e* his greater sensitiveness is indicated by a more animate Mount of Venus. (See plate II.)

III. The Hands of a Boy of Sixteen

The Line of Fate is lacking in the left hand, the Head Line is short and has a hump below the middle finger. The Life Line is unusually broad and undifferentiated. The Mount of Venus is strongly emphasised and large, but the rest of the hand is without animation.

The boy has criminal tendencies and was repeatedly convicted for theft. His criminality is chirologically expressed by the absence of the Line of Fate, which suggests a lack of social adjustment, the broad Life Line and the large Mount of Venus, which indicates the predominance of the instincts, and an undifferentiated vitality, and the short Head Line, which is the sign of a fairly poor intellectual capacity and, as is revealed by its hump, of mendacity. Brute vitality combined with a lack of social adaptability and a mediocre brain dispose the boy to crime. (See plate III.)

IV. The Hands of a Boy of Fourteen

The right hand shows only a fragmentary section of the Line of Fate, which is completely absent in the left. The Head Line of the left hand is interrupted. The Heart Line ends on a particularly large Mount of Jupiter. Line *e* indicates great sensitiveness.

The boy is mentally backward and a violent stutterer. The stuttering is suggested by the discrepancy between the interrupted Head Line and the high degree of ambitiousness indicated by the Heart Line proceeding straight to the prominent Mount of Jupiter, and the sensitiveness of the boy. (See plate IV.)

V. The Hand of a Boy of Fourteen

All the main lines appear in the hands of this boy. He has good dispositions but fear and intricate influences of the milieu prevent their proper development. The fact that the Line of Fate (*b*) springs from the Mount of the Moon shows a certain amount of self-reliance and opposition to the milieu. The Line of Milieu, (*g*) prominently outlined, represents the inhibiting influences of the milieu. This line emerging from the Pre-natal Line (*f*) suggests that a pre-natal influence is still effective in the milieu. The Head Line (*c*) proceeding in a good swinging movement across the palm reveals a scientific turn of mind, the hump in it an inclination to mendacity which in this particular case is probably conditioned by fear. The well-curved Heart Line (*d*) designates good-naturedness, and the branch which extends from it to the Head Line reveals that the intellect strongly influences the emotions.

The mother of the boy had experienced a severe shock in her pregnancy because of the unfaithfulness of her husband. The boy was always very much afraid of his father, though treated by him with kindness. It was impossible for him to develop properly in the atmosphere of his parental home. The well-designed Line of Fate (*b*) and the good Head Line showed, however, that he was capable of development, and indeed as soon as the boy was sent to a boarding-school he began to develop according to his natural dispositions. (See plate V.)

VI. The Hand of a Boy of Twelve

The appearance of the hand of this boy, who is the brother of the youngster described in the preceding analysis, is perfectly normal. There is neither a Line of Milieu nor a Pre-natal Line, and all the lines appear in their proper relationship. The boy is a normal child without fears and inhibitions and normally gifted.

Though the mother in the period of her second pregnancy had again discovered that her husband had been unfaithful to her, she was less affected than the first time. Consequently the boy was not pre-natally influenced, and altogether was of a more harmonious and stronger disposition than his more unstable brother. (See plate VI.)

VII. The Hands of a Boy of Six

This hand, which has an abundance of lines for a boy of six, shows a very unstable, irritable and highly strung child. The right hand, which also conveys the impression of restlessness because of its too many lines, indicates that the instability of the child is due to heredity. The Head Line of the right hand shows a strong inclination to moods of depression. The vivid Mount of Venus and the double Girdle of Venus of the left hand suggests considerable sensuality and a premature development in this direction. The forking of the end portion of the Head Line denotes a critical mind, the prong declining towards the Mount of the Moon a tendency to depression, and the little hump in the Head Line below the second finger a mendacious disposition. The combination of the critical faculty and mendaciousness manifests itself as cunning craftiness. The somewhat short Heart Line proceeding to the Mount of Jupiter signifies despotism and self-interest. The feature in the Zone of Milieu points to a severe traumatic incident in early childhood ; the sudden death of his mother when he was two.

Summing up, we can say that this boy of six is very unstable, prematurely developed, intelligent, cunning, grasping, somewhat melancholy. and extremely sensual. He has hysterical fits of vomiting. (See plate VII.)

VIII. The Hands of a Woman of Forty

These prints, which illustrate the preceding case, are the hands of the aunt of the boy, the sister of his mother. They show that the melancholy disposition is hereditary in the family of the boy, for the Head Line in both hands of his aunt is forked early and its downward bend shows a very dejected mind. (See plate VIII.)

IX. The Hands of a Girl of Three

This print is the chirological portrait of an egoistic, despotic, militant, energetic, independent, but also intelligent child, though its intelligence is of a more material kind. The hand as a whole suggests great vitality and strength, combined with a certain instinctiveness and impulsiveness. Her egoism and desire to dominate are betrayed by the straight Line of Heart leading to the Mount of Jupiter. Her fighting spirit is indicated by a line springing from the Mount of Mars (*x*), her energy and independence by the strong Line of Fate proceeding from the Mount of the Moon, the good, somewhat materialist, intellectual capacity by the straight Head Line which drops only slightly towards its end. The materialist tendency is emphasised by the low position of the Heart Line, and her vitality and strength are clearly expressed by the forceful Mount of Venus and the already very pronounced Line of Fate. (See plate IX.)

X. The Hand of a Boy of Four

The unusual formation of the Heart Line which cuts straight across the hand indicates a violent and uncontrolled child. The much-torn Line of Fate denotes irritability and lack of concentration. The short and strangely formed Head Line, with a square downward on it, reveals a peculiarity of the nervous system.

The boy has an unusually violent temper, he is erratic, desultory, lacking in concentration, very irritable and susceptible to a kind of epileptic fit. (See plate X.)

XI. The Hand of a Boy of Fifteen

This hand indicates a very energetic, self-reliant, intelligent boy who has great possibilities of development. His energy and self-reliance can be deduced from the Line of Fate which, though starting late, comes from outside and is powerfully developed.

His good intelligence is denoted by the well-defined Head Line. Its incisiveness, in contrast to the otherwise simple lining of the hand, indicates a more practical kind of intelligence. The chainlike design of the Head Line suggests a good combinative ability. The Line of Milieu (G) designates milieu difficulties; the Line of Fate coming from outside (as already mentioned), shows his independence and a development that is in contrast to that of his milieu.

The boy is of very humble origin and his family lived in great poverty, which was the cause of his milieu difficulties. He left his home, however, in his boyhood, and already at the age of thirteen played a leading part in a youth movement. Thus the milieu difficulties acted as a stimulus in a positive direction. (See plate XI.)

XII. The Hands of a Girl of Thirteen

In this hand the Line of Milieu is almost as strong as the delicate and cut-up Line of Fate. The Head Line clearly reveals a tendency to depression and is forked early. The decidedly chainlike pattern of the Heart Line is the indication of a nervous heart. The right hand is very much like the left because it has also a weak Line of Fate, a Head Line expressive of melancholy and a chainlike Heart Line.

The negative and inhibiting milieu influence emanates from the delicate mother who, subject to melancholic moods, had a most dispiriting effect on the child. The influence is very active because the girl herself is very delicate, uncertain and wavering, as suggested by the disjointed Line of Fate, and highly strung and dejected, as indicated by the early forked and drooping Head Line. There is little hope of development and improvement as the similarity of the personal-individual and the parental hand reveals the compulsion of heredity. (See plate XII.)

XIII. The Hands of a Boy of Thirteen.

The hand shows a boy capable of development in scientific thought, but he is strongly dominated by his instincts and a sexuality which borders on the abnormal. His fear of life is connected with complex and strong influences of the milieu. The right hand shows the predominance of the instincts, lack of control and also powerful sexual desires.

The possibilities of development can be deduced from the

straight line of Fate coming from the Mount of the Moon, his scientific ability from the Head Line terminating below the fourth finger, the fear of life from the Line of Anxiety on the Mount of Venus, the sexual anomaly from the squares on the Mount of Venus, the strong assertion of the instinctive from the broad Line of Life, and a prominent Mount of Venus and the influences of the milieu from the numerous small lines and the Line of Milieu.

The married life of the parents is very disrupted and the father is a sadist. The unhappy marriage and the sadism of the father have caused great suffering to the boy, who himself showed sadistic inclinations already in his early childhood, but as he is intelligent and capable of development a psycho-therapeutic treatment could do a great deal of good. (See plate XIII.)

XIV–XVI. The Hands of a Boy and His Two Sisters

These three plates of the left hands of three children are to demonstrate the difference of Head Lines.

Plate XIV shows the long but rather rigid Head Line of a boy of thirteen which suggests a moderate degree of intelligence without much differentiation. The boy is the least intelligent of the three.

The Head Line of his younger sister (plate XV), who is nine years old, is not so rigid and more alive, indicating a more scientific ability.

The youngest (plate XVI), a girl of seven, has the most mobile and liveliest Head Line, though it forms a little hump below the Mount of Saturn. It is of a chainlike pattern and forked at its end. The child is the most intelligent of the three and has a very nimble, critical mind, though she is somewhat insincere. (See plates XIV–XVI.)

XVII–XVIII. The Hands of a Girl at the Ages of Twelve and Nineteen

The fact that hands change continuously is most strikingly illustrated in this case. The hand of the girl of twelve (plate XVII) shows only the four main lines and a Line of Technical Intuition, in fact it is a normal hand with all its possibilities of development.

At the age of nineteen the picture has completely changed (plate XVIII) and the hand is that of a fully developed, grown-up person. Apart from the former lines, everything is new : the Line of Intuition, the influences of the milieu, the animate Mount of the Moon, the Girdle of Venus and the various lines of experience. In fact, many decisive experiences and changes took place in the milieu of the girl between her twelfth and nineteenth years, which explain the influence of the milieu on the hand and her considerable development, for when she was twelve a series of conflicts started between her parents which came to such a head when she was seventeen that a divorce, in the same year, became inevitable. (See plates XVII and XVIII.)

XIX–XX. The Hands of a Boy at the Ages of Twelve and Eighteen

This comparison of two plates from the same person at different times is intended to demonstrate once more the changing of the lines of the hand. At the age of twelve (plate XIX) the Line of Fate is still delicate and undefined, and six years later (plate XX) it is much stronger and above all much clearer in the space between the Heart Line and Head Line. In the eighteenth year the Line of Intuition which springs from the Line of Fate is much more developed, and finally we find an entirely new line which, rising between the Mounts of Venus and Moon, envelops the latter. (See plates XIX and XX.)

XXI–XXII. The Hands of Two Fraternal Twins at the Ages of Twelve [1]

The left hands of the boys are different, particularly the two Head Lines. The Head Line in plate XXI of the twin who was born one and a half hours earlier is forked, the one prong going on to the Mount of Mars, whereas the other is slightly drooping. The Head Line of his brother (plate XXII) is not split but definitely declining to the Mount of the Moon thus indicating depression. The entire aspect of the hand in plate XXI is less lively than that of plate XXII. The Heart Lines are very similar and both finely curved. As is often the case with twins, the younger (plate XXII) is more sensitive, unstable and clumsier, more inclined to meditation. They are

[1] Modern biology distinguishes two kinds of twins : *Identical Twins*, which are always of the same sex and have many characteristics in common. They originate from one ovum or egg fertilized by one sperm, whereas *Fraternal Twins* have their origin in two eggs or ova each fertilized by a separate sperm. They need not be of the same sex, and are as different as ordinary brothers and sisters.—(*Translator's Note.*)

both very good-natured and warm-hearted. (See plates XXI and XXII.)

XXIII. The Left Hand of the Brother of the Twins, aged Seventeen

This plate is intended to show the great difference between the hand of the elder brother of the twins and their hands. It is much less lively and has a clearly outlined Girdle of Venus. He is more passive and more difficult than his twin brothers. (See plate XXIII.)

APPENDIX TWO

THE HANDS OF THE MENTALLY DISEASED

Experience has shown that the study of the psychology of the abnormal mind has yielded valuable knowledge of the workings of the normal mind, neurotic and otherwise. The same approach can be used in Chirology. Apart from widening its area of research it would help to systematize and clarify a branch of human knowledge which still harbours a number of vague ideas and unprecise conceptions.

From the material I collected I have chosen the prints of forty-six hands belonging to a group of schizophrenics and a group of non-schizophrenic, mixed cases. This comparatively small number is hardly more than a drop in the ocean and too limited to allow of valid deductions, for to make those at least several hundreds of cases would have to be studied and analysed, and made available. This would be an enterprise worthy of some scientifically minded chirologists of any school, for though their analyses may differ, the facts, data and prints collected by them would be the same.

The compilation of material as offered in this Appendix is by no means an easy task, and sometimes even the fundamental requirements of the chirologist are not understood. I need only mention the case of one hospital where I applied for permission to carry out research work. The reply was that it would be no use going there, because in their institution not only the minds of the patients were deformed but also their hands.

The actual taking of prints of the hands of mental patients is often fraught with difficulties because many patients have to be coaxed into opening their hands, especially those whose general attitude is negative, defensive or hostile. In certain cases prints could be obtained only with the assistance of the nursing staff. A number of patients whose hands were chirologically extremely interesting had to be left out altogether. These difficulties of a physical and technical nature provide an additional reason for co-operation among chirologists who are interested in this particular field of research.

In all these cases the outer hand is of utmost importance. It is a pity that in Chirology, so far as I know, no systematic use has yet been made of photography complementary to prints. Although Spier himself did not work with photographs, I have always felt the need for it, especially in his particular teaching which attributes such great importance to the outer hand, and the raised and dropped positions. The palm, after all, tells only half the story. The technical and financial problems are of course the main obstacle. To make up for the lack of photos in this edition the reader will find short notes on relevant points relating to each case. On the whole these notes are not characterological analyses, but are of a chirological, descriptive nature.

Before analysing these hands one might have thought that certain chirological features which frequently appear in the hands of neurotics would also occur in the hands of abnormal people; and yet, in a number of cases they are most signally absent. I am referring to the hollow palm and the Line of Anxiety. A description of the hand in its dropped position, when not lying flat on the table, will be found on page 36 and an illustration in fig. 42. This feature is accentuated if the palm is hollow. It is surprising that this should not occur in twelve out of twenty-eight schizophrenic cases; the palms of their hands are not hollow, and in some cases extraordinarily flat, but only two of them (plates XLI and XLII) were able to drop their hands. Incidentally, both have properly developed Lines of Fate in their left hands.

The Line of Anxiety, (see p. 73)[1] which one might expect to find in the hand of modern man almost as a matter of course, proves to to be absent in twenty-three of the fifty-six hands of the schizophrenics. In contrast to this only four hands in thirty-six of the mixed group are entirely without it, while a further three hands seem to have had such a line which now appears to be in a state of deterioration. A tentative explanation may be that the non-schizophrenics are still able to worry about their condition, whilst the schizophrenics have ceased to do so being no longer conscious of theirs.

The nails, too, deserve our attention, but I am not prepared to say whether it is of any significance that eleven of the twenty-eight schizophrenic cases have long, curved nails, which shows a

[1] The Anxiety Line, the Line of Prevention (in this context perhaps more correctly termed Hypochondriac Line), and the Double Life Line are of the same kind expressing various degrees of intensity. Therefore the reference "No Anxiety Line" on the two tables always refers to this whole group of lines.

disposition to diseases of the lung. In an additional six cases the nails are also curved, although of mixed shapes. The moons of the nails (see p. 46) are of some interest, for of the twenty-eight schizophrenic cases fourteen have no moons at all, not even on their thumbs, and an additional seven have none except on their thumbs. Strikingly ribbed nails are found at the ratio of ten in the twenty-eight of the schizophrenic group with an additional five not so heavily ribbed. This constitutes a marked difference to the non-schizophrenic group where this is not an outstanding feature.

An interesting point is, perhaps, the occurrence of the outward bent will-phalange of the thumb (see p. 22) of cases LIV, LVI and LVII (one manic-depressive, two mania cases).

In contrast to the schizophrenic group the non-schizophrenic, mixed group has no difficulty in performing the dropped position of the hands. (Exceptions for pathological reasons were cases LV and LVII.)

Another feature worthy of out attention is connected with the Head Line. In twenty of the fifty-six hands of schizophrenics the Head Line begins at a distance from the end of the Life Line (see p. 72 and fig. 67) in contrast to the thirty-six hands of non-schizophrenics in which these two lines join.

The Thyroid Lines appear in the space between the Heart Line and the fingers (see plate LXIV, left), *i.e.* the space of the Girdle of Venus and the Ring of Saturn. They consist of a number of small lines forming a girdle. Further study will be necessary more precisely to distinguish between a Double Girdle of Venus (or the more frequent appearance of a half or third of such a girdle) and the Thyroid Lines, and their possible chirological interconnection. Unfortunately, like human beings, the lines of the hands do not always oblige us by presenting clear-cut test cases, and the Subsidiary Lines (see pp. 91–96) offer added difficulties, for they are usually in a state of flux. Incidentally, this is the main reason why it is chirologically impossible to make a prognosis. Only a diagnosis may be ventured on the strength of chirological symptoms of past and present. Bearing this in mind, the vexed question of prognosis should never arise.

As far as Fate Lines are concerned, we find in the schizophrenic group that in twenty-four of fifty-six hands they are either missing, fragmentary, or very poorly developed. In the non-schizophrenic, mixed group this amounts to eleven of thirty-six hands. In the schizophrenic group there is no instance in which both hands of the same case are without Fate Line. In the mixed group the Line of Fate is missing in both hands in plates LIII and LIV. The rest of the Lines of Fate can be divided into groups of faint, frayed, interrupted, "islanded" lines, and some that begin very late (see plate XXVIII left), or are inclined towards the Finger of Intuition (plate XXXII left), or are less strongly developed than the Line of Intuition (plate XXIX left). In contrast to this, however, we find some well-developed Fate Lines which, paradoxically, may be in keeping with the patients' particular affliction. The reader will find these cases listed in the two tables.

Are we justified in seeing a certain affinity between plates XXIV and XXV; XXVI and XXVII; XXVIII, XXIX, and XXX? (Incidentally, this last group of three catatonic schizophrenic cases could be linked with the print of a catatonic schizophrenic in Charlotte Wolff's book, *The Human Hand.*) Studying the prints of two manic-depressive cases (plates LIV and LV) we may notice that the spheres of the Heart and Head Lines are tending towards each other.

Finally a few notes on the rather important feature of consistency (p. 16). Considering that softness of consistency indicates an adaptable, flexible personality, but also instability and inconstancy, our findings in this respect are perhaps not surprising. Of fifty-four hands of schizophrenics (Case XXXVI could not be investigated because the patient was incapable of relaxing the muscles of her hands) forty-seven had soft or very soft consistency, five medium-soft and one pair of hands medium-hard (Case XXXVIII). Of the mixed group of thirty hands (the consistency of the hands of Cases LII, LXI and LXVII was not established) twenty-three proved to be of soft or very soft, three of medium-soft, three of medium-hard (LXII left and LXV left and right) and one of hard consistency.

This short survey must not be concluded without a word about a problem which seems to preoccupy many people: the borderline between normality and insanity. To a certain extent many of us are very abnormal at times and perhaps a little abnormal all the time, for we are all made of the same stuff; and yet, we do not become inmates of an asylum because personal effort allows us to mould that stuff we are made of. In many instances the step from sanity to insanity is a rung on the same ladder, offering the possibility of going up or down. Therefore it stands to reason that chirological features which appear in the hands of the cases described in this Appendix are often found in those of normal persons. It is, therefore, impossible to diagnose dormant abnormality or approaching insanity. Our destiny will always *lie in our hands* not as a static, irrevocable threat, but rather like a mould which shapes and reshapes itself according to what we pour in.

From all this I hope one point emerges clearly: chirology is in need of more people with a medical background who are prepared to devote time and considerable energy to serious study. Without this Chirology cannot hope to lay claim to scientific recognition. Even if these notes produce no other result than this conviction, they will not have been written in vain.

NOTES ON HANDS OF SCHIZOPHRENICS (PLATES XXIV–LI)

XXIV. The Hands of a Case of Primary Dementia

Male, age forty-six. Nails: broad. Consistency: left, soft; right, medium-soft. Raised position: changing from one extreme to the other, fingers and thumbs clenched, alternating with opening of hands the will-phalanges of the thumbs pointing away from the index fingers, a phenomenon frequently found in normal adolescents. Very broad lines. Right and left hands similar.

XXV. The Hands of a Case of Simple Schizophrenia

Male, single, age eighteen. Nails: lung, but not curved on thumbs and indexes; no moons except on thumbs. Consistency: left, soft; right, medium-soft. Thumbs bent in raised and dropped position. Broad lines. Empty mounts. Right hand: Fate Line missing. Left hand: Head Line broken, first section to Heart Line, second takes the place of a forceful Line of Experience swooping down towards Mount of Moon.

XXVI. The Hands of a Case of Primary Dementia

Male, age twenty-nine. Nails: throat, curved, some ribs; no moons except on thumbs and index of the right hand. Consistency: left, soft; right, medium-soft. Raised position: thumbs away from indexes, but will-phalanges bent. Can drop hands without difficulty. Left hand: no Fate Line, Thyroid Lines.

XXVII. The Hands of a Case of Primary Dementia

Female, age thirty-eight. (Psycho-motor retardation.) Nails: broad, curved, some ribs; no moons. Consistency: soft. Colour of hands: bluish-red. Raised position: hands inclined inward with left thumb pointing outward and right thumb upper phalange inward (may be due to some injury at ball of thumb). Empty Mount of Venus. No Lines of Experience. Head Line begins away from Heart Line. Similarity of right and left hands.

XXVIII. The Hands of a Catatonic Schizophrenic

Male, single, age thirty-three. Nails: broad; no moons. Flat palms. Consistency: soft. Dropped position: after initial difficulties palms lie quite flat on table, with thumbs straight and pointing away from indexes. Right hand practically no Fate Line. Empty mounts in both hands. Broad lines.

XXIX. The Hands of a Catatonic Schizophrenic

Male, age thirty-seven. Nails: no moons. Flat palms. Consistency: soft. Thick, undifferentiated fingers, bulky upper phalanges. Deep-set thumbs (see p. 22 and fig. 24) with bulbous will-phalanges (especially right hand) stand upright in raised position; left ring finger has prominent outward bend (in contrast to neurosis cases who usually hide this finger below the middle finger). This agrees with the marked Line of Intuition, which is stronger than the Fate Line, in the left hand. Dropping of hands demands a tremendous, almost terrifying effort; patient lifts them like a heavy weight; exhausted by the physical exertion drops them stiffly like two blocks of stone. Right hand: no Fate Line, no Lines of Experience, empty mounts.

XXX. The Hands of a Catatonic Schizophrenic

Male, single, age twenty-two. Nails: no moons. Consistency: soft. Thick upper phalanges. Thumbs: particularly prominent will-phalanges, but bent in raised position. Finds dropping of hands difficult. Mount of Venus: horizontal lines. Fairly empty mounts except that of Venus.

XXXI. The Hands of a Schizophrenic

Female, single, age fifty. Nails: lung, some ribs; moons on all, only on thumb nails small and faint. Consistency: very soft. Raised position: thumbs bent and pressed to indexes. Dropped position: difficult, dropped her hands just once; the completely curled-in fingers touch the table with the backs of first phalanges. Left hand: two Heart Lines, Ring of Saturn, connecting line from Head to Heart Line. Right hand: Girdle of Venus, Thyroid Lines.

XXXII. The Hands of a Schizophrenic

Female, single, age sixty-five. Nails: curved horizontally only; no moons except on thumbs; much ribbed. Flat palms. Consistency: very soft. Raised position: left thumb stands up independently. Incapable of dropping hands. Several significant chirological symptoms appear in both hands: two Head Lines;

prominent connecting line from Head to Heart Line springing from island formation; main Head Line begins at slight distance from Life Line.

XXXIII. The Hands of a Schizophrenic

Female, single. Nails: lung, and more than usually curved; no moons; ribs on thumbs only. Flat palms. Consistency: very soft. No mount elevations in lower hand region. Incapable of dropping hands. Right hand: two Head Lines, one with island, the other two-pronged and descending into the Mount of the Moon region.

XXXIV. The Hands of a Schizophrenic

Female, single, age thirty-seven. Nails: bronchial, some ribs; no moons. Flat palms. Consistency: very soft. Raised position, left hand: little finger stands away, ring finger curled in, rigid thumb points outward. Incapable of dropping hands. Two Life Lines in right hand, broken Life Line in left.

XXXV. The Hands of a Schizophrenic

Female, married, age forty-four. Nails: lung; no moons. Colour of hands: blue. Consistency: medium-soft. Raised position: thumbs with humps (see p. 20 and fig. 15) upright without angle to indexes. Incapable of dropping hands. Left hand: no Fate Line, Thyroid Lines.

XXXVI. The Hands of a Schizophrenic

Female, single, age twenty-three. Nails: lung; moons on all fingers though faint on right thumb; apart from thumbs all nails are blue in moon region with adjoining white and then blue patches. Colour of hands not affected. Incapable of dropping hands. Fragmentary Line of Fate in either hand. No Lines of Experience. Left hand: very short Head Line; empty Mounts of Venus, Mars and Moon. The taking of prints was difficult as she kept her hands clenched.

XXXVII. The Hands of a Schizophrenic

Female, single, age forty-three. Nails: heart, much ribbed; no moons. Flat palms. Consistency: very soft. Incapable of dropping hands. In left hand no Fate Line (although small fragments can be seen). In both hands the broken Head Lines emerge at a distance from the Life Lines and in either case the first section runs into the Heart Line. A possibility of Thyroid Lines.

XXXVIII. The Hands of a Schizophrenic

(Schizophrenic excitement with marked confusion.) Female, married, age forty-three. Totally deaf. Nails: healthy looking; no moons except on thumbs. Consistency: medium-hard. Raised position: fingers clenched, thumbs bent. Dropped position: same. No Fate Line in right hand.

XXXIX. The Hands of a Schizophrenic

Female, age forty. Nails: lung, ribbed; no moons except very faint on right thumb. Flat palms. Consistency: very soft. Colour of hands: reddish-blue. Incapable of dropping hands. In both hands the Head Line begins away from the Life Line.

XL. The Hands of a Case of Early Schizophrenia

Boy, age fifteen. Mother, single, served a prison sentence for neglect of her children. (Brother affected too.) Nails: long, not curved, with a few ribs; moons on all fingers. Hollow palms. Consistency: very soft. Raised position: the balls of the thumbs protruding, will-phalanges bent. Can drop hands without difficulty. Mount elevations of Venus and Moon absent. Right hand: hardly any Fate Line. Left hand: prenatal Line; small Milieu Line with square; Fate Line in sections beginning at some distance from it. In both hands Head Line begins at considerable distance from Life Line.

XLI. The Hands of a Schizophrenic

Male, age thirty-four. Nails: long, broad, curved, blue tinge. Flat palms. Consistency: soft. Raised position: balls of thumbs protruding. Can drop hands without difficulty. Developed Fate Line in both hands.

XLII. The Hands of a Schizophrenic

Female, married, age fifty-five. Nails: mixed shapes, much ribbed; no moons except on thumbs. Flat palms. Consistency: left, soft; right, very soft. Large hands for her size. Can drop them without difficulty. Developed Fate Line in left hand.

XLIII. The Hands of a Schizophrenic

Female, age thirty-four. Violent, impulsive, unco-operative. Prints could only be taken with help of two nurses. Nails: lung. Flat palms. Consistency: very soft. Dropping of hands impossible. Developed Fate Line in left hand. In both hands Head Line begins at distance from Life Line.

XLIV. The Hands of a Schizophrenic

Female, married, age thirty-one. Nails: lung, much ribbed; no moons except on thumbs. Flat palms. Consistency: soft. Incapable of dropping her hands. Left hand: Mount of Venus horizontal lines. Both hands: undeveloped Fate Line; Head Line begins at distance from Life Line; no Lines of Experience. Right and left hands similar.

XLV. The Hands of a Schizophrenic

Female. Nails: particularly small and broad on little fingers; no moons. Consistency: very soft. Raised position: left ring finger, complete inward bend. Dropped position: all fingers curled in, thumbs bent. Head Line begins away from Life Line in right hand but not in left. No Lines of Experience in either hand.

XLVI. The Hands of a Schizophrenic

Female, married, age forty-five. Nails: lung, ribbed; no moons. Consistency: soft. Raised position: will-phalange of left thumb points outward. Incapable of dropping her hands. Lower Mounts of Venus empty. Frayed main lines. Thyroid Lines.

XLVII. The Hands of a Schizophrenic

Female, single, age fifty-four. Nails: unusual consistency; long, curved sideways, strongly ribbed and dotted; moons only on thumbs; dusty grey shade. Colour of hands: normal. Flat palms. Consistency: soft in rather mountless hands. Incapable of dropping them. Frayed main lines. Milieu Lines. Prenatal Lines. Girdle of Venus. Important chirological features similar in both hands.

XLVIII. The Hands of a Schizophrenic

Female, age forty-four. Nails: lung, ribbed; no moons; yellow. Colour of hands: blue. Consistency: very soft. Raised position: left ring finger curls in. Incapable of dropping her hands. Uncooperative and antagonistic; very difficult to take prints. Left hand: overcast with lines; prominent Line of Intuition, which might develop into a Barring Line.[1] Right hand: Line of Fate with two large islands.

XLIX. The Hands of a Schizophrenic

Female, married, age fifty-two. Nails: lung, ribbed; no moons. Consistency: soft. Incapable of dropping her hands. Right hand. Barring Line. Left hand: potential Barring Line.

L. The Hands of a Schizophrenic

Female, single, age thirty-seven. Nails: long, curved; no moons. Consistency: very soft. Incapable of dropping her hands. Thyroid Lines.

LI. The Hands of a Schizophrenic

Female, age twenty-six. Nails: ribbed; no moons except faint ones on thumbs. Consistency: very soft. Raised position: backs of hands stand up straight, fingers fall inside. Dropped position: hands lie one in front of the other, fingers pointing in opposite directions, with the lower edge of the palms touching the table. Both hands: Mount of Venus horizontal lines; Head Line begins rather far from Life Line.

[1] Spier called a *Barring Line* a line which runs in a swoop from the Mount of the Moon to the Mount of Mercury, expressing a barrier effect between the subconscious and the conscious mind.

NOTES ON HANDS OF NON-SCHIZOPHRENICS (MIXED) (PLATES LII–LXIX)

LII. The Hands of a Paranoiac

Male, married, age fifty-one. Although suffering from hallucinations and delusions, and convinced that he is remotely controlled by rays, he can talk quite intelligently and with a fine command of language. Right hand: fragments of a Fate Line. Left hand: gap in Fate Line between the age of approximately eight to twenty-two. Small islands appear on the Head Line. In both hands the Mount of Venus is covered with mainly horizontal lines.

LIII. The Hands of a Paraphreniac

Female, married, age thirty-five. Sudden onrush of illness, excitable, hallucinations. Nails: bronchial; no moons except on thumbs; healthy looking. Consistency: soft. Raised position: thumbs stand away from indexes with will-phalanges bent. Dropped position: thumbs relaxed and straight. Both hands: no Fate Line; horizontal lines on Mount of Venus. Left hand: small Head Line; particularly large space between Heart Line and fingers.

LIV. The Hands of a Case of Manic-Depressive Psychosis

Male, married, age sixty-two. Deluded. Believes he has contracted syphilis and therefore imagines he will lose his sight. He is interested in music and very proud of his piano playing. States he has hundreds of friends and no enemies. Nails: long, curved in left hand; no moons except on thumbs. Consistency: left, soft; right, medium-soft. Raised position: hands stiff, will-phalange of thumbs bent very much outward. Dropped position: hands fall on to fingertips only, palms, wrists and arms held up in the air. Both hands: Fate Line missing; connecting line from Head to Heart Line; both these lines close together, hence large space between Heart Line and fingers. Right hand: Mount of Venus horizontal lines, and bare section; no Lines of Experience. Left hand: Ring of Saturn.

LV. The Hands of a Case of Manic-Depressive Psychosis

Male, married, age sixty-eight. Catatonic kind of limb stiffness. (Maintaining any limb in the position in which it is placed.) He told me, he was "wrong in the head", tried to commit suicide, and had no will power. Nails: very broad and flat; large moons on both thumbs, none on two fingers of each hand, small moons on others. Consistency: soft. Incapable of raised and dropped position. Both hands: Heart Line, with island, cutting right across palm; no Head Line. Left hand: very prominent, negative Line of Experience almost as if taking up place of Head Line.

LVI. The Hands of a Case of Mania

Female, married, age fifty-seven. Nails: bronchial, ribs; no moons except on thumbs. Consistency: left, soft; right, medium-soft. Raised position: thumbs very much bent outward. Dropped position: table is touched with the tips of outstretched fingers only, back of hands forming a triangle with fingers; thumbs lie straight and relaxed (a feat usually not met in cases of neurosis). Both hands: two Life Lines. Right and left hands similar.

LVII. The Hands of a Case of Mania

Male, married, age sixty-seven. Shell-shock case, 1914–18 war. Mutters incoherently to himself, but when questioned gives clear and reasonable answers. Draws imaginary figures in the air with his hands. Nails: lung, ribbed; very large moons on seven fingers. Consistency: soft. Thumbs are bent outward. No chirological observations on dropped hands possible because of excessive trembling. Both hands: Head Line swings to Heart Line; very shallow lines as if fading out. Right and left hands similar.

LVIII. The Hands of a Case of Psychoneurotic Depression

Female, married, age twenty-three. (Her sister also in a mental hospital.) Nails: bronchial; moons: right hand, none except on thumb; left hand, none on ring and little fingers. Consistency: soft. Raised position: thumbs straight though very close to indexes. Drops hands in a jerky manner, left thumb lies straight, right upper thumb phalange slightly bent. Both hands: Prenatal Lines. Left hand: hardly any Head Line. Right hand: unusually late starting Fate Line.

LIX. A Case of Uncertain Diagnosis

Melancholia? Schizophrenia? Conversion-Hysteria? Female, single, age forty-five. Food faddism, tube-fed daily. Had times when she could not bear to be touched. When taken out for walks the act to turn round and return home became a harassing symptom. Delusions that someone was going to shoot her. Utters sentences like: "I do not want tea because I would not love. I would not live." At the same time, however, as if she belonged to a distant world where different ideas and standards prevail, she would smile resignedly at us "stupid, ignorant" people, who would not understand. Nails: bronchial, ribbed and curved; no moons; colour: normal. Colour of hands: very white. Consistency: left, very soft; right, soft. Raised position: thumbs, particularly in left hand, stand up straight with lower phalange bone protruding. Dropped position: ball of hands and fingertips touch table; thumbs are bent. Beautiful, raphaelesque expression of hands. In both hands large island on Head Line. Left hand: two branches of Life Line; well-developed Fate Line.

LX. A Case of Uncertain Diagnosis

Paranoid Schizophrenia? Psychoneurosis? Female, married. Worried about her heart. Father and daughter suffered from heart trouble. Although at some time her own heart sounds were rather faint, no abnormal developments have set in. Nails: heart; no moons except on thumbs. Very frail-looking upper phalanges. Consistency: soft. Raised position: thumbs stand well. Left hand: slightly interrupted Head Line with island; interrupted Life Line. Both hands: Ring of Saturn; Line of Prevention (or Hypochondriac Line).

LXI. A Case of Uncertain Diagnosis

Schizophrenia or Thyroid Illness? Female, single, age twenty-two. (Brother in hospital too.) Very happy child until she developed peculiar moods. (Mother has had some thyroid trouble.) Relaxed, loose dropping of hands. Indication of Thyroid Lines.

LXII. A Doubtful Case

Male, single, age twenty-five. (Sister has thyroid trouble.) Nails: short; no moons except on thumbs. Consistency: right, hard; left, medium-hard. (The only hard consistency of all the hands I examined at the hospital.) Raised position: hands fall towards himself, thumbs bent. Dropped position: hands, pointing towards each other, resting only on fingertips, backs of hands pointing outward, fingers inward, both little fingers striving away from ring fingers, in left hand more than in right. Very large hands compared with his size. Right hand: Girdle of Venus. Left hand: Thyroid Lines.

LXIII. The Hands of a Case of Bromide Intoxication

Male, married, age thirty-one. Nails: broad, flat; no moons on both little fingers. Consistency: soft. Dropped position: fingers curled in. Left hand: drooping Head Line. Both hands: Mount of Venus, horizontal lines; Thyroid Lines.

LXIV. The Hands of a Case of Reactive Depression. (Caused partly by Thyro-Toxiosis.)

Female, married, age forty-nine. Thyroid trouble of long standing, had throat operation in connection with it. Nails: healthy looking; no moons except small ones on thumbs. Consistency: soft. Right hand: two islands on Head Line. Left hand: prominent Line of Intuition. Thyroid Lines.

LXV. The Hands of a Case of Involutionary Melancholia

Female, married, age sixty-four. Involutionary Melancholia with mild degree of agitation. Depressed and hypochondriacal, but no conduct disorders or delusions. Her intellect and her sense of orientation are functioning well. Nails: no moons. Consistency: medium-hard. Raised position: thumbs straight, all fingers curl in. Dropped position: patient either drops hands stiffly with a bang or very slowly and anxiously. Both hands: Line of Prevention (or Hypochondriac Line.)

LXVI. The Hands of a Case of Involutionary Melancholia

Female, single, age forty-nine. Foreign housemaid. Confused and depressed. Nails: mixed, curved shapes; no moons except on thumbs. Consistency: soft. Raised position: hands relaxed, thumbs bent. Dropped position: hands stiff and clumsy, thumbs straight. Both hands: Anxiety Line, possibly developing into a Line of Prevention (or Hypochondriac Line). Left hand: islanded Poison Line (see p. 92).

LXVII. The Hands of a Case of Involutionary Melancholia

Female, single, age sixty-two. Maternal aunt and one sister committed suicide. Invalid mother was cared for by patient until patient was forty years old. Depended a great deal on her mother in making decisions. Left hand: prominent drooping Line of Experience part of which, together with a Gastro-enteric Line, forms a Barring Line. Right hand: Prenatal Line extending into Milieu Line.

LXVIII. The Hands of a Case of Involutionary Melancholia

Female, married, age fifty-seven. Nails: bronchial, many ribs; no moons except on thumbs. Consistency: very soft. Raised position: left thumb straight with lower phalange protruding, right thumb straight but close to fingers. All fingers of both hands stiffly pressed together. Both hands: connecting line from Head to Heart Line.

LXIX. The Hands of a Case of Senile Dementia

Male, married, age seventy. Nails: long, broad and curved with ribs; large though faint moons on all fingers and thumbs. Consistency: left, soft; right, medium. Raised position: stiff, thumbs very much bent outward. Dropped position: only the tips of fingers and thumbs touch the table. Both hands: Heart Line straight and in sphere of Head Line. Undeveloped Fate Line. Large island formation connects Life and Head Lines. Similarity of right and left hands.

TABLE ONE

SCHIZOPHRENIC GROUP

Abbreviations: L=left; R=right; H=hand. The arabic figures of Tables One and Two refer to the plates.

Nails, lung	L & R	25, 31, 33, 35, 36, 39, 43, 44, 46, 48, 49
—, bronchial	L & R	34
—, curved (not lung)	L & R	26, 27, 32, 41, 47, 50
—, heart	L & R	37
—, many ribs	L & R	32, 37, 39, 42, 44, 46, 47, 48, 49, 51
—, some ribs	L & R	26, 27, 31, 34, 40
—, no moons	L & R	27, 28, 29, 30, 33, 34, 35, 37, 39, 45, 46, 48, 49, 50
—, no moons exc. thumb	L & R	25, 32, 38, 42, 44, 47, 51
Colour of hands or nails blue/bluish-red	L & R	27, 35, 36, 39, 41, 48
Flat palms	L & R	28, 29, 32, 33, 34, 37, 39, 41, 42, 43, 44, 47
Consistency, soft		24 L, 25 L, 26 L, 27 L & R, 28 L & R, 29 L & R, 30 L & R, 31 L & R, 32 L & R, 33 L & R, 34 L & R, 37 L & R, 39 L & R, 40 L & R, 41 L & R, 42 L & R, 43 L & R, 44 L & R, 45 L & R, 46 L & R, 47 L & R, 48 L & R, 49 L & R, 50 L & R, 51 L & R
—, medium		24 R, 25 R, 26 R, 35 L & R, 38 L & R
—, hard		None
Dropped Position, difficult or incapable of	L & R	28, 29, 30, 31, 32, 33, 34, 35, 36, 37, 39, 43, 44, 46, 47, 48, 49, 50
Two Life Lines		34 R
Two Head Lines		32 L & R, 33 R
Two Heart Lines		31 L
Missing Head Line		None
Fate Line, missing		25 R, 26 L, 29 R, 35 L, 37 L, 38 R
— —, undeveloped		24 R, 27 L & R, 28 R, 32 R, 33 L & R, 34 L, 36 L & R, 37 R, 40 L & R, 44 L & R, 46 R, 47 L, 48 L
— —, well-developed L.H.		41, 42, 43
Frayed Main Lines		46 L & R, 47 L & R
Milieu Line		27 R, 31 L, 33 L, 34 L, 35 L, 40 L, 46 L & R, 47 L & R
Island on Head Line		32 L & R, 33 L & R, 34 L & R, 35 L, 41 R, 49 L
Distance from Life to Head Line		27 L & R, 32 L & R, 33 L, 37 L & R, 39 L & R, 40 L & R, 42 R, 43 L & R, 44 L & R, 45 R, 50 R, 51 L & R
Mount of Venus, empty		25 L & R, 27 L & R, 28 L & R, 29 R, 36 L & R
— — —, horizontal lines		30 L & R, 44 L, 51 L & R
No Experience Lines		27 L & R, 28 R, 29 R, 36 L & R, 44 L & R, 45 L & R
No Anxiety Line		24 L & R, 27 L & R, 29 R, 31 L, 32 L & R, 33 L & R, 34 L, 35 L & R, 36 L, 38 L, 40 L & R, 46 R, 47 L & R, 50 L, 51 L & R
Prenatal Line		25 L & R, 26 L, 31 L & R, 37 L, 40 L, 41 L, 47 L & R
Barring Line		49 R
Thyroid Lines		26 L, 31 R, 35 L, 37 L & R, 46 L & R, 49 R, 50 L & R
Similarity of right and left hands		24, 27, 32, 44

TABLE TWO

NON-SCHIZOPHRENIC (MIXED) GROUP

Nails, lung	L & R	57
—, bronchial	L & R	53, 56, 58, 59, 68
—, heart	L & R	60
—, flat	L & R	55, 63
—, many ribs	L & R	68
—, some ribs	L & R	56, 57, 59, 69
—, no moons	L & R	59, 65
—, no moons exc. thumb	L & R	53, 54, 56, 60, 62, 64, 66, 68
Consistency, soft		53 L & R, 54 L, 55 L & R, 56 L, 57 L & R, 58 L & R, 59 L & R, 60 L & R, 63 L & R, 64 L & R, 66 L & R, 68 L & R, 69 L
—, medium		54 R, 56 R, 62 L, 65 L & R, 69 R
—, hard		62 R
Raised Position, thumb straight or outward	L & R	54, 56, 57, 58, 59, 60, 65, 68, 69
Dropped Position, thumb straight	L & R	53, 54, 56, 66, 69
Dropped Position, difficult or incapable of	L & R	55, 57
Two Life Lines		56 L & R, 59 L
Missing Head Line		55 L & R
Hardly any Head Line		53 L, 58 L, 69 L & R
Fate Line, missing		53 L & R, 54 L & R
— —, undeveloped		52 R, 55 R, 57 L & R, 66 L, 69 L & R
— —, well-developed	L.H.	55, 56, 58, 59, 64, 65, 67, 68
Milieu Line		59 R, 64 L, 67 R, 68 L
Island on Head Line		52 L, 57 L, 59 L & R, 60 L, 64 L & R, 65 R, 69 L & R
Distance from Life to Head Line		None
Mount of Venus, empty		69 L
— — —, horizontal lines		52 L & R, 53 L & R, 54 R, 63 L & R, 66 L & R
No Experience Lines		54 R
Anxiety line probably deteriorated		55 L, 57 L, 62 L
No Anxiety Line		64 R, 68 L, 69 L & R
Prenatal Line		58 L & R, 66 R, 67 R, 68 L
Barring Line		67 L
Thyroid Lines		61 L & R, 62 L, 63 L & R, 64 L
Similarity of right and left hands		56, 57, 69

POSTSCRIPT

As a rule, a personal word of the author precedes his work and, therefore, is usually skipped by the eager reader, who wishes to get to grips with the book itself. In trying to foil this attempt of the reader by disguising my preface as a postscript, I am solely guided by the wish that the following lines should not suffer an equal fate.

Not only this book, but in fact my entire chirological activity, owes everything to the lively personal interest Professor Dr. C. G. Jung has shown in my work. It was Professor Jung who many years ago encouraged me to make the interpretation of the hand, then a self-taught hobby, my profession and the work of my life, and he also helped me with valuable advice during the period of my professional readjustment. He never failed to lend me support on my sometimes difficult path, and finally bore witness to his sympathetic interest in my work by writing the introduction to this, my first publication in book form. I, therefore, wish this book to be the expression of my gratitude to Jung, without whose encouragement and support my work would not have met with the response which stimulated me to go on with my research.

Professor Dr. H. W. Maier of the Kantonalen Heilanstalt (Asylum) " Burghoelzli " very kindly provided me with several opportunities of testing my experience, gained up to then in my private practice only, in his Institute. Though the material available was limited it helped me to consolidate the foundations of my method.

Last, but not least, I must thank Dr. Bernhard Diebold, who was the first to write about my work in a well-informed and comprehensive article, published in 1929 in the *Frankfurter Zeitung*, thus arousing interest for chirology among a section of the public which so far had looked upon it with scepticism and disapproval.

During the last ten years I have been working hard to perfect my chirological method, and to put its foundations, which up to that time had been chiefly intuitive, on a scientific basis. I had to go through a series of positive and negative experiences before I was able to submit the basic principles of my method to the public. My range of experience has been considerably enlarged by all those to whom chirology means more than fortune-telling, who, therefore, acquainted themselves with its problems and tried to deal with them in an objective manner. Among these I must, above all, count my pupils and the host of people who sought my advice. My gratitude goes out to them as well.

In this general distribution of thanks I should hate to omit those who out of fear of compromising themselves or out of inveterate scepticism have refused to associate themselves in any way with chirology and have not even stooped to give it the chance of a critical examination, categorically declaring that it was all so utterly unscientific. Their negative attitude has never failed to stimulate me to subject my work and my findings to ever fresh examination and adjustment.

I may be permitted to point out once more that it is not the purpose of this book to prove the solely scientific character of chirology but to show its manifold possibilities with special emphasis—as is indicated by the title—upon that which I believe to be the most essential task of the chirologist : to assist the growth and the development of personality in the child.

If I have succeeded in this book to stimulate an interest in chirology, particularly in people whose duty it is to guide and further the development of children, I shall have the satisfaction of having redeemed a fraction of the gratitude I owe to those who made it possible for me to write this work.

J. S.

INDEX

This Index which I have compiled for the student of chirology is intended to be three things: an index, a brief and concise extract of the book, and a questionnaire which will enable the student to find out to what an extent he has mastered the meaning of chirological features. Numbers in *italics* indicate a main point of reference.—THE TRANSLATOR.

The subject matter of Appendix Two is not included in this Index because it is sufficiently covered by Tables One and Two.

Abdominal diseases, 45
abnormality, constitutional, 49
accidents, 70, 85
action, 58, 74
active type, 14, 17, 74
actors, hands of, 13, 62
adaptability, 16, 24, 25, 26, 74, 75, 76
—, bad, 33, 39, 40, 42, 82, 88
adjustment, lack of social, 159, 160
Adler, Gerhard, 63
æsthetic exaggeration in sexual matters, 91
— interests, 13, 62
aggressiveness, 63
alcohol, 27, 49
alimentary functions, 45
ambitiousness, 23, 61, 74, 76, 84, 91, 94, 160
amiability, 60, 65
anæmia, 45
analytic treatment, 102–104
ancestors, 75
ancestral hand (right h.), 2, 5, *47–57*, 68, 107
— tradition, 55
angle, of thumb, 20, 21
animality, 14
animals, love of, 60, 64
animateness, 28
anomalies, 12
—, incurable, 49, 75
—, sexual, 164
antagonism, 52
Anxiety, Line of, 73, 81, 164
Apollo, Mount of, 62, 104, 105
—, Mounts of and Mercury, combined, 105
architects, hands of, 104
architecture, talent for, 31
Aristotle, 4
artistic talents or interests, 13, 28, 29, 60, 62, 65, 87, 104
artists, hands of, 14, 33, 35, 62, 76
ascending lines, definition of, 69
— — from Mount of Moon, 96
asocial attitude, 21, 60, 76, 88
association, power of, 82
astrology, xi
avarice, 61

Back of hand, 36, 74
— — —, delicate, 92
— — —, large, bulky, 14, *15*
— — — and fingers, 37
balance, inner, 82
bashfulness, 32
beauty, sense of, 13, 62
benevolence, 84
Beziehungen zwischen dem Ich und dem Unbewussten, 63
blackmailers, hands of, 65
bladder, 33, 93
blood, 17, 23, 45, 46
— pressure, 45
bodily sphere, 15
bottled-up development, 80
brain diseases, organic, 80
bronchia, 44
bronchial catarrh, 44
brothel proprietors, hands of, 65
brutality, 17, 22, 83

Cabaret artists, hands of, 13
calculative type, 84, 85, 86
callousness, 83, 84
calumniators, hands of, 65
capriciousness, 108
cardiac diseases, 30, 44, 84, 86
Carus, Gustave, 4
central nervous system, 67, 80
change, important, 72, 75
— of profession, 75
changeability, 46, 63
changing of lines, 164–165
character, instability of, 16
child, normal, 161
childhood illnesses, 71
—, impressions, 70
—, paradise, 161
—, suppression in, 76, 77

childhood, hands of new-born, 68
—, Head line, 108
—, intuition of, 29
— Line of Fate, 106, 107
—, lines and subsidiary lines, 68, 69, 106, 108 ; maze of lines, 161
—, lines of experience, 106
—, love of, 60
—, maturity of lines, 106
—, pampered and spoilt, 107
—, time conception of, 72
—, urge for power, 107
chiro-analyst, his personality, 7, 98 ; his qualifications, 8, *98–99* ; his task, 2, 7, 57, 66
chirology, advantage of ch. method, 98
—, the learning of, 97
—, scientific approach, xi, xii, 66
—, therapeutic importance, 2
chiromancy, xi, 4, 47
circulation of blood, 17, 45, 46
circulatory disturbances, 45
clairvoyance, 65, 88
claustrophobia, 26, 37
clumsiness, 14, 165
coldness, 84, 85, 86
collectors, hands of, 23
collective consciousness, 98, 99
— knowledge, 11
— unconscious, 63, 71, 88, 99, 103
colour of hands, 17
— of nails, 45
combination, power of, 82, 163
Combinative Line of Intuition, 88
commercial talent, 62
composers, hands of, 104
composition of blood, 23, 45, 46
—, talent for, 31
concentration, difficulties of, 20, 108, 162
conductors, hands of, 104
conflict, professional, 38
conical hand, *13*, 16, 17, 81
— little finger, 30
consciousness, 27, 30, 38, 40, 63, 81, 99
consistency, of hands, 16, 17
—, lack of, 13
constancy, 65
constipation, 63
constitution, weak, 73
control, lack of, 162, 163
convention, breaking away from, 55
conventional attitude, 21
craftiness, 161
cranks, 88
creativeness, 14, 29, 59, 100, 104
criminality, 20, 22, 65, 75, 159, 160
criticism, 46, 81, 161, 164
—, benvolent, 83
—, constructive, 83
cunning, 161
curiosity, 40

d'Albert, Eugen, 104
dancers, hands of, 13
death, 70, 96, 161
degeneration, 81
dejection, 37, 55, 162
delicate work, suitability for, 15
depression, fits and periods of, 33, 49, 56, 81, 82, 83, 101, 161, 163, 165
descending lines, definition of, 69
despotism, 161, 162
desultoriness, 71
detail, neglect of, 16
determination, lack of, 13
development, belated, 108
—, difficulties of, 107, 108
—, progressive, 95
—, standstill of, 108
Diebold, B., 167
diet, 27
" disappointed " people, 28
discontent, 101
diseases, abdominal, 45
— of the bladder, 83, 93
— of the brain, 80
—, cardiac, 30, 44, 84, 86
—, disposition to, 18
—, infectious, 93
— of the liver, 27, 86, 93
—, organic, 44
—, pulmonary, 11, 45, 46, 73
disharmony, 12
displeasure, 101, 107
dispute, seeking, 89
distances between fingers, 20, 24, 25, 31, 32, 40, 41, 102, 105
distrust, 42
doctors, hands of, 14, 62
domination, 26, 94, 162
dreams, 103
drug addicts, 64, 92
drunkenness, 64
duality of personality, 82
duty, sense of, 41
dynamic aspect of hand, 5, 12, 43
— development, 11
dynamism of personality, 66

Education, 29
ego and the *id*, 33
egocentric attitude, 36, 37, 39, 98, 99
egoism, 14, 86, 162
embitterment, personal, 99, 107
emotional coldness, 28, 86
— element, 13, 15, 46, 95
— excitability, 87
— -intuitive forces, 41
— response, 84
— sphere, 84
— uncertainty, 41
— zone, 103
emotion and reason, wavering between, 84
—, control of, 85
—, undifferentiated, 85
emotions influenced by intellect, 84, 160
energetic type, 162
energy, arrested, 23
—, frittering away of, 20, 108
—, psychical, 20
Engelhard, R., 5
enjoyment, life of, 61
Entdeckung der Seele, Die, 63
enteric illnesses, 34, 93
enthusiasm, 15
environment, 11, 17, 24, 25, 28, 35, 36, 37, 40, 59, 66, 74, 75, 76, 77, 78, 86, 95, 101, 107
epilepsy, 162
erotic instincts, 65
erotically responsive, 42
erratic type, 85, 162
escapism, 50, 100
excitability, 46, 63, 86, 87
excitable people, 17
Experience, Lines of, *95–96*, 104, 165
experiences, negative, 56
—, the necessity of, 99
exploitation, 62, 65
external world, 23, 62, 96, 100
extraversion, 13, 19, 36, 37, 55, 62, 91, 92
— of parents, 56
Extraversion, Intuitive Line of, 88

Failure, fear of, 20
family feeling, 61
fanaticism, 64
Fate, Line of, 70, *73–78*, 79, 88, 89, 93, 106, 107
Fate, Line of, absence of, 75, 76, 159, 160
—, — —, chronology of, 78
—, — —, course of, 74
—, — —, cut-up, 162, 163
—, — —, ending at Head line, 74
—, — —, ending at Index, 76
—, — —, fragmentary, 160
—, — —, interrupted, 75
—, — —, island on, 94
—, — —, relative strength of, 80
—, — —, square on, 94
—, — —, starting at Life line, 76
—, — —, starting at Mount of Moon, 76, 160, 162, 164
—, — —, starting in Mount of Moon, 77
—, — —, straight, 74, 164
—, — —, strong, 74, 162
—, — —, threadlike, 75
—, — —, weak, 74, 81, 163
father fixation, 32, 51, 54
father *imago*, 75
fatigue, 45
fear, 78, 80, 93
fear of life, 25, 37, 50, 73, 163
— of responsibility, 32, 42
feeblemindedness, 159
feeling, 15, 39
fighting spirit, *62–63*, 162
fingers, 15, *18–34*
— changing shape, 38
— closely held together, 41
—, flexible, 105
— inclined to unconscious sphere, 102
—, inward bend, 37, 40
—, long, 15
—, long, and delicate back, 92
—, long and thin, 102
—, outward bend, 37
fixation, 17
flaccidity, 70
flattery, open to, 62
flexibility, 13, 16, 17, 21
—, lack of, 22
forcefulness, 74
form, sense of, 30
fortune-telling, 4, 6
frankness, 21
friction, seeking, 89
frivolity, 65

Gall, 27, 86, 93
gardening, 108
gastric illnesses, 93

Gastro-enteric Line, 93
— — in bits, 93
— —, straight and pale, 93
generation, deadlock of, 49
—, problem of, 53–56
—, task of, 55
generosity, 21, 84
Girdle of Venus, 91, 92, 161, 165, 166
glandular trouble, 46
Goclenius, Dr., xi
goodnaturedness, 85, 160, 166
gout, 17, 27
grammatical-linguistic gift, 31
Greeks, ancient, 59
guilt, feeling of, 50, 82, 90

Hand, expression of, 10, 11
—, inner, 11, 43
—, outer, 12, 43
handicraft, 108
hands, xi
—, colour of, 17
—, dropped, 35, 36, 43
—, fleshy, 61
—, intellectual-spiritual, 62
—, large, 15
—, reddish, 64
—, shape of, *see* conical, philosophical, primitive, rectangular, spatular
—, small, 15
—, uplifted, 35, 36
hardness, 76
harmony, 15, 19, 24, 28, 36, 74, 106, 161
—, lack of, 62, 64, 76
head injuries, 80
Head line, 77, *80–84*, 85, 86, 108, 162, 163
— —, bad, 83
— —, branch to beginning of Heart line, 85
— —, branching off to Heart line and Mount of Moon, 83
— —, branching off to Mount of Moon, 81, 161, 162
— —, broad and straight (barrier effect), 82
— —, chainlike, 82, 163, 164
— —, chronology, 83–84
— —, difference of Head lines, 164
— —, double, 82
— —, drooping to Mount of Moon, 81, 103, 165
— —, ending before middle finger, 80
Head line, ending below little finger, 80, 105, 164
— —, — — middle finger, 80
— —, — — ring finger, 80
— —, forked below middle finger, 81, 108
— —, — — ring finger, 81, 83
— —, — to Mounts of Mars and Moon, 83, 165
— —, good, 83
— — and Heart line as one line, 85
— —, hump on, 83, 159, 160, 161, 164
— —, incisive, 163
— —, interrupted, 83, 160
— —, island on, 94
— —, missing, 159
— —, rigid, 164
— —, short, 159
— —, square on, 162
— —, steep drop to the Mount of Moon, 83
— —, thin, 82
hearing, acute sense of, 43
heart, 30, 84, 163
—, illnesses of, 30, 44, 84, 86
—, nervous, 163
—, small, 84
Heart line, 83, *84–86*, 161
— —, branch to beginning of Head line, 84
— —, — to Mount Jupiter or Head line, 84
— —, chainlike, 86, 163
— —, chronology of, 86
— —, course of, 84
— —, dark colouring of, 86
— —, descending lines on, 86
— —, ending before middle finger, 84
— —, — on Mount Jupiter, 84, 160, 161
— — and Head line as one line, 85
— —, island on, 86, 94
— —, low position of, 162
— —, smaller branches to Head line, 84
— —, space above, 86
— —, — between Head line and, 86, 165
— —, springing from Head line, 84
— —, straight, 83
— —, — across (barrier effect), 84, 162

Heart line, straight, ending on Mount Jupiter, 84, 162
— —, triangle on, 95
— —, unusually red, 86
— —, well-curved, 160
heredity, 12, 48, 52, 63, 75, 106, 108, 161, 162, 163
highly strung child, 161
— — persons, 71, 91
homicidal type, 22
homosexuality, 54, 91
humanity, love of, 60
humility, lack of, 100
humps on thumb, 20
hyper-critical intellect, 81
hyper-intellectual people, 103
hyper-sensitiveness, 25, 72
— in sexual matters, 91
— of vascular nerves, 46
— to noise, 43
hypochondria, 73, 79
hysteria, 63, 161

Identification, psychological, 50, 73
idiocy, 76
idlers, asocial, 88
illnesses, 94, 96
—, bladder, 33, 93
—, disposition to, 18
—, effect of, 70, 71, 72
—, enteric, 34
—, hypochondriac fear of, 79
—, kidney, 33, 93
—, liver, 27, 86, 93
—, uterine, 33
imagination, 15, 39, 46, 63, 64
imago, 32, 48, 75
immovability, 16
impetuosity, 46
impotence, psychical, 43, 75
imprisonment, 94
impulsiveness, 19, 33, 72, 162
incalculable, 71
incautious, 85
inconstancy, 16, 65
independence, 20, 21, 162
Index, *23–27*, 61, 91
—, bend sideways, 24
—, bend sideways to middle finger, 27, 40
—, changing shape, 38
—, conical, 24
—, crooked, 38
—, inward bend, 25
—, length of, 26
Index, long, 26
— and middle finger tending to ring finger and little finger, 40
—, outward bend, 24, 25, 27, 38, 40
—, rectangular, 24
—, slightly rectangular, 39
—, smaller than ring finger, 26
—, straight, 24, 40
— —, and knotty, 40
—, thickening at lowest phalange, 27
—, turning away, 39, 40, 41, 42
Indians, hands of, 58
inertia, 70
infantilism, 49, 100
infectious diseases, 93
inferiority feelings, 21, 41, 55, 71, 80, 91, 100
— —, overcoming of, 100, 103
— —, various kinds of, 101
—, organic, 101
inhibition, 21, 22, 37, 41, 42, 43, 56, 72, 76, 77, 80, 89, 103, 107, 160
—, lack of, 85
inner world, 24, 87
insatiableness, 31
inscrutability, 43
insincerity, 17, 82, 83
instability, 16, 22, 63, 65, 161, 165
instincts, 64, 71, 73, 85, 159, 162, 163
intellect, 15, 33, 34, 39, 41, 74, 80, 81, 84, 85, 89
— influenced by emotion, 85
—, poor, 34, 76, 80, 160
—, rigid, 82
—, undifferentiated, 85
Intellect, Finger of (middle finger), 90
intellectual-spiritual sphere, 37
intelligence, 162, 163
—, practical, 15
internal secretion, disturbances of, 86
intestines, 34
intolerance, 101
introversion, 19, 27, 28, 36, 37, 74, 91
introvert, creative, 14
intuition, xi, xii, 24, 28, *29*, 66, 67, 68, 80, *87–89*, 93, 165
Intuition, Finger of (Ring finger), 90
Intuition, Lines of, 29, 78, 79, *87–89*
Intuition, Line of, 87, 104, 105
— — —, beginning at Line of Fate (combinative), 88, 165
—, — —, — at Life line (productive), 87
—, — —, — at Mount of Moon (imaginative), 88

Intuition, Line of, ending between ring finger and little finger (technical), 87, 105, 164
—, — —, — on ring finger (artistic), 87
—, —, triangle on, 95
intuitive element, 13
— —, inhibited, 29, 103
— —, manifested, 89
— —, strengthening of, 89–90
— —, weakened by narcotics, 93
irrational attitude, 36, 46, 81, 88
irresponsibility, 37
irritability, 161, 162
Islands, 94–95
— on Head line, 94
— on Heart line, 86, 94
— on Life line, 94
— on Line of Fate, 94
— on Poison line, 92
isolation, 42

Jealousy, 42
joints, knotty, 14, 17
Jung, C. G., xi, xii, 8, 13, 19, 25, 35, 48, 63, 64, 102, 167
Jupiter, Mount of, 61, 84, 85, 91, 161, 162
—, absence of Mount of, 61
—, lines on, 61
—, prominent, 61, 160
—, squares on, 94
—, too large, 61
juridical trend, 80
justice, sense of, 40

Kidneys, 33, 93
kleptomania, 22
knots on joints, 17
knuckles, 36

Laotse, 25
Lavater, 4
laziness, 17
libido, transferred, 50, 60
life-force, 18, 71
Life line, *69–73*, 78, 79, 93, 96
— —, bad, 70
— —, chronology, 72
— — crossing Head line, 72
— — ending at Head line, 72
— —, frayed, 71
— —, good, 70
— —, interrupted, 71
Life line, island on, 94
— —, not joining Head line, 72
— —, square on, 94
— —, starting point, *69–70*, 71
— —, too broad, 70, 159, 164
— —, triangle on, 95
lines, *see* Life line, Lines of Fate, of Milieu, of Prevention, of Anxiety, of Luck, of Experience, Pre-natal line, Gastro-enteric line, Poison line
—, absence of main, 49, 159
—, disappearing of, 67
— of intuition, 29 ; *see* intuition
Lines of Travel, 64
—, terminology of, 68
linguistic talent, 30
Little finger, *30–33*, 38, 43, 91
— —, clinging to hand, 32
— —, conical, 30, 42, 105
— —, crooked, 45, 64
— —, deeply rooted, 42
— —, inward bend, 32, 38
— —, jutting out, *31–32*, 42, 43
— —, knotty lower phalange of, 42
— —, long, 31
— —, normal length, 30
— —, outward bend of tip, 43
— —, rectangular tip, 31, 104
— —, small rectangular, 31
— —, spatular, 104
— —, straight and long, 105
— —, — and rectangular, 31
— —, without moon, 33
liver, 27, 86, 93
logicality, 33
Luck, Line of, 74
lunacy, 49
lung diseases, 11, 46, 73

Maier, H. W., 167
Mars, Mount of, *62–63*, 65, 89, 96, 162, 165
—, — —, large, 62
—, — —, normal size, 62
—, — —, without lines, 63
masculinity in women, 78, 79
materialism, 65, 82, 85, 102, 162
materialistic expression of hand, 27, 62
materiality, 39, 55
mediator, 33
medical talent, 60, 105
meditative mind, 165
mediumistic knowledge (conscious), 42
— powers, 41

melancholia, 39, 82, 162, 163
—, pathological, 83
memory, 82
mendacity, 62, 82, 83, 102, 160, 161
mental deficiency, 76, 159
mentally backward, 160
Mercury, Mount of, 62, 65, 93
— and Mount of Apollo combined, 105
—, too strong, 62
methodical mind, 13, 33
Middle finger, 27, *33–34*, 38, 73, 74
— —, bend to Index, 40
— —, crooked, 34
— —, inclined to Ring finger, 41
— —, long, 33
— —, long and thin, 61
— —, outward bend, 38
— —, — — of upper phalange, 41
— —, rectangular, 33, 105
— —, thickened lowest phalange, 34
— —, tip bent to Ring finger, 34
— —, too long, 90
— —, too short, 33
milieu, 11, 35, 37, 75, 76, 77, 78, 96, 160, 163, 165
—, fear of, 160, 163
— fixation, 35, 36, 50, *51*, 52, 53, 89, 90, 101
— —, definition of, 51
Milieu, Line of, *77–80*, 95, 163
—, — —, continuous, 77
—, — —, relative strength of, 80
—, — —, small lines, 78
Milieu, Marks of, 78, 164
—, zone of, 161
misanthropy, 61
mission, personal, 54, 55, 56
moodiness, 101
Moon, Mount of the, 35, 59, 61, *63–64*, 71, 76, 77, 78, 83, 92, 96, 160, 162, 164, 165
— — —, animate, 165
— — —, ascending lines on lower part of, 63
— — —, big, without lines, 64
— — —, cross lines on upper part, 64
— — —, normal size and few lines, 64
— — —, with maze of lines, 103
moons, 30, 33, 46
—, absence of, 45
mother fixation, 32, 42, 51, 53, 54, 67
— *imago*, 75
motherliness, 60
Mounts, *see* Apollo, Jupiter, Mars, Moon, Neptune, Saturn, Venus
—, absence of, 58, 59
—, consistency of, 58
—, size, 58, 65
—, terminology, 59
movability, 17
musicality, 30, 65, 87, 105
musicians, fingers of, 38, 104
—, hands of, 62, 104

Nail, pale-coloured, of middle finger, 27
—, colour of, 45
—, dots on, 45
—, ribbed, 45
nails, 44–46
narcissism, 39
narcotics, 93, *see* drug addicts, poison line
nature, connectedness with, 64
negativism, 29
negotiating ability, 31
Neptune, Mount of, 65
nervous constitution, 41, 81
— disturbances, 82, 83, 94, 108
— systems, 67
— —, defects of, 80
neurosis, 74, 91, 103
neurotic symptoms, 26, 37, 52
nursing, gift for, 60

Obduracy, 16
observation, power of, 24, 38, 39, 40, 89
obstinacy, 22, 64
optical sensibility, 40
optimism, 37, 46
oratory, gift for, 30
orderliness, 17
organising ability, 26, 39, 61
originality, 14
outward world, 23, 62, 96, 100
ovaries, 33
overcompensation, 71, 91
over-dramatising, 98
over-sensitiveness, 23
overstrain, 20, 81

Painters, hands of, 88
Paracelsus, 4
parent-fixation, 23, 51, 101
parents, 3, 9, 50
—, difficulties of, 107
—, false ambitions of, 107, 108

parents, guidance of, 109
—, imperfections of, 101, *106–109*
—, opposition to, 52
—, psycho-therapeutic treatment for, 107
—, unsolved problems of, *53–56*, 106, 107
partner-relationship, 30, 32, 42, 43, 50, 53, 54, 61, 95, 104
passiveness, 17, 58, 74, 166
pathological cases, 12
patient, the psychology of the, 6, 7, 54, 55, *98–104*
peasants, hands of, 14
pedagogical talent, 60, 61, 95
pedantry, 17
persona, 35, 74, 89, 99
personality, split of, 83
—, synthesis of, 1
phalanges, 14, 22, *39–43*
—, linking of, 20
phantasy, 15, 64, 81
pharynx, 45
philosophical hand, 14
— tendency, 61
phonetical-linguistic talent, 31
physical inferiority, 71
— power, 18, 70, 71, 162
— weakness, 20
Physiognomy, 4
pigheadedness, 22
Pindar, 1
pink hand with greyish tinge, 17
Poison line, 92–93
— —, island on, 92
poison, sensitiveness to, 92
political talent, 62
position of hand, raised, 35
power, craving for, 71, 84
practical sphere, over-emphasised, 37, 40
practicality, 13
precocity, 81, 108
prediction, 4, 6, 7
prejudice, 107
Pre-natal line, 79, 160
Prevention, line of, 73
pride, 23
primitive hand, *14*, 62, 159
primitiveness, 49, 102
Productive line of Intuition, 87
prudishness, 32
psycho-chirologist, *see* chiro-analyst
psycho-chirology, *see* chirology
psycho-pathic case, 49, 75
psycho-therapeutic treatment, 21, 41, 53, 56, 67, 80, 103
Psychologische Typen, 25
publicity, desire for, 55, 62
pugnacity, 62

Quarrelsome type, 46, 62, 89

Rape, 94
rationality, 16, 39, 46, 74, 80, 81, 90
realisation, inner, 100
realism, 16
reality, sense of, 15, 27
—, lack of sense of, 88
realities, world of, 28, 100
reason, 13, 16, 19
recklessness, 33, 37
recognition, desire for, 85
rectangular hand, 12, *13*, 17, 81, 105
red hand, 17
reflection, 19
reliability, 13
religiosity, 61, 89, 100
repression, 60, 73, 75, 78
resignation, vanquished, 92
resistance passive, 63
—, power of, 95
responsibility, fear of, 32
restlessness, 20, 161
" re-valuation of values ", 53
rheumatism, 27, 45
right hand, 2, 5, *47–57*, 68, 107
rigidness, 16
Ring finger, *28–30*, 38, 86
— —, bend sideways to Index, 40
— —, — — to middle finger, 28, 40, 41
— —, crooked, 30
— —, inward bend, 28, 29, 38
— —, length of, 28
— —, long, 28
— —, outward bend, 42
— —, straight and spatulate, 29, 105
Ring of Saturn, 91

Sadism, 164
Saturn, Mount of, *61–62*, 65, 164
—, — —, absence of, 61
—, — —, large, 61
—, Ring of, 91
Savonarola, 4
scepticism, 13, 46
schoolmistresses, 26
sciatica, 27

scientific ability, 62, 80, 105, 160, 163, 164
— method, 67
scientists attitude to chirology, 59, 66
—, hands of, 14, 62, 76
scruples, lack of, 65
sculptors, hands of, 55, 104
Secret of the Golden Flower, 63
secretion, internal, 86
secretory disturbances, 45, 84
self-assertion, 18, 19, 24, 26, 39, 61, 71, 91
self-assuredness, 39, 101
—, development of, 89, 101
—, lack of, 76, 82, 90
self-confidence, strengthening of, 89, 90, 98, 105
self-control, lack of, 63, 87
self-criticism, 83, 88
self-interest, 161
self-reliance, 20, 39, 90, 160
—, lack of, 21
self-sufficiency, 21
self-torment, 61
self-willed persons, 46
selfishness, 85
sense-perception, 24, 38
sensibility, 42
sensitiveness, line of, 159, 160
— to poison, 92
sensitivity, 24, 61, 159, 165
sensuality, 31, 59, 161
seriousness, 65
sexual life, 30, 31, 32, 33, 60, 91
sexuality, abnormal, 163
—, callous, 60
—, powerful, 163
—, premature, 161
—, undifferentiated, 60
shape of fingers, changing, 38
shyness, 32
sight, sense of, 24
sixth sense, 65
skilful hands, 15
sluggishness, 37
smell, sense of, 24
sober-mindedness, 13, 16, 81
social adaptability, 26
— —, bad, 39, 40
sociality, 25
spatular hand, 14
speech, gift for, 30
Spier, Julius, ix, xii
spirituality, 10, 15, 27, 39, 55, 78, 87, 91, 92
spirituality, over-emphasised, 37
spleen, 27
spontaneity, 15, 19
sport, interest in, 62
squares, 94–95
—, large and small, 94
— on Fate line, 94
— on Life line, 94
— on Mount of Jupiter, 94
— on Mount of Venus, 94
—, relative importance of, 95
stability, 17
stage fright, 93
stammering, 26
staying power, 17, 19
static aspect of hand, 5
— element, 11
stomach, 27
story-tellers, hands of good, 88
strain, 45
stuttering, 37, 160
subconscious, 103
sublimation, 60, 73, 91
subsidiary lines, *91–93*, *95–96*
success, seeking of, 55, 62
suppleness, 21
susceptibility, 22
suspicion, 13, 46
symbolism, of earth, the finite, heaven, the infinite, 15
Symbolism of the Human Form, 4
symbolism of numbers, 13
synthesis of personality, 1
systematical mind, 13

Tact, lack of, 21
talents, suppressed, 108
Tao Teking, 25
taste, sense of, 24
technical ability, 105
Technical Line of Intuition, 87, 105
telepathic powers, 65
temper, uncontrollable, 22
tenacity, 16, 22
tension, inner, 40
terminology, 8
—, mythological, 59
testicles, 33
thumb, *18–23*, 40, 74
—, angle of, 20, 21
—, broken appearance, 21
—, changing shape, 38
—, clinging to hand, 39
—, club-shaped, 22
—, flexible, 30, 105

thumb, forceful, 13, 21
—, good, 26, 70
—, long, powerful, 19
—, phalanges, *19–22*, 26, 39
—, rigid, 39
—, set-in, 33
—, shape of, 18
—, short, 14
—, small, 20
—, stiff, 22
—, too small, 39
—, weak, 75, 81
thyroid, glands, 46
tonsils, 45
touch, sense of, 24
touchiness, suppression of, 72
tradition, break in, 55, 56
trauma, 79, 94
traumatic shock, 78, 79, 94, 160, 161
Travel, Lines of, 64
triangles, 95
—, on Heart line, Life line, Line of Intuition, 95
twins, 165
tyranny, 76

Üeber die Energetik der Seele, 64
Ular, Alexander, 25
uncertainty, 20, 39, 42, 107
unconscious, 28, 34, 41, *63–64*, 71, 78, 81, 82, 84, 87, 88, 89, 96, 102, 103, 107
— attitude, 35, 38
—, collective, 63, 71, 88, 99, 103
— emotional life, 77
— transfer of impulses, 60
— violence, 22
uncontrollable child, 162
unconventional attitude, 20, 41
unevenness, 20
unintelligent persons, 102
unreliability, 82
unscrupulous behaviour, 76
uterus, 33

Vanity, 61, 107
" vanquished resignation ", 92
vascular nerves, 46
— trouble, 45, 86
Vedic period, 58
vegetative nervous system, 67
Venus, Girdle of, 91, 92, 165, 166
—, Double Girdle of, 161
—, Mount of, 22, 23, 31, 32, 35, 43, *59–61*, 62, 65, 69, 70, 71, 72, 73, 74, 79, 91, 94, 95, 96, 159
—, — —, absence of, 60
—, — —, absence of lines on, 60
—, — —, big and strong, 60, 64, 159, 162
—, — —, horizontal lines on, 60, 61
—, — —, lower part of, 59, 60
—, — —, normal size, 60
—, — —, square on, 94, 164
—, — —, upper section, 59, 60, 105
—, — —, vivid, 161
—, — —, weak, 60
violence, 22
violent people, 17, 162
— temper, 162
vitality, 18, 39, 70, 71, 162
—, impaired, 72
—, poor, 20
—, undifferentiated, 160
vivacity, 19, 20
vocal ability, 30
vocational guidance, 70, 104
vomiting, fits of, 161
vulnerability, 72

Warmheartedness, 60, 84, 166
white hands, 17, 92
will-power, 18, 19
—, poor, 75
workers, hands of manual, 67, 68
working men, hands of, 14
World-finger (Index), 23
wounded, hands of, 48

Yoga, 58

INDEX OF FIGURES

FIG.	PAGE
1	10, 11, 28
2	10, 11
3	12, 13
4	12, 13
5	12, 13, 15
6	12, 13, 15
7	12, 14
8	14
9	12
10	14
11	16
12	17
13	19
14	19
15	20
16	20
17	21
18	21
19	21
20	21
21	22
22	22
23	22
24	22
25	24, 27, 43
26	24, 40
27	25, 40
28	25
29	27
30	29, 30, 32
31	31, 42
32	31
33	31
34	32
35	32
36	33
37	33
38	33
39	35
40	35, 36
41	35
42	35, 36
43	36, 41
44	36
45	37, 42, 43
46	37
47	38
48	39
49	40
50	40
51	41
52	41
53	41
54	41
55	42
56	44
57	44
58	44
59	44
60	44
61	44
62	45
63	39, 59, 60, 61, 62, 63, 65
64	60, 61
65	63
66	69, 73, 77, 79, 80, 84, 87, 92, 93
67	71, 72, 75, 80
68	71, 76, 80, 83, 84, 93
69	72, 80, 84, 91, 93
70	72, 81, 84, 95
71	72, 79, 84, 86
72	73, 75, 81, 86, 93
73	64, 77, 81, 86, 88, 91, 94, 95
74	81, 84, 91, 96
75	82, 83
76	82, 86, 95
77	82, 86
78	85
79	84
80	83
81	83, 87
82	83
83	83
84	83
85	84, 85
86	85
87	86

PLATE I

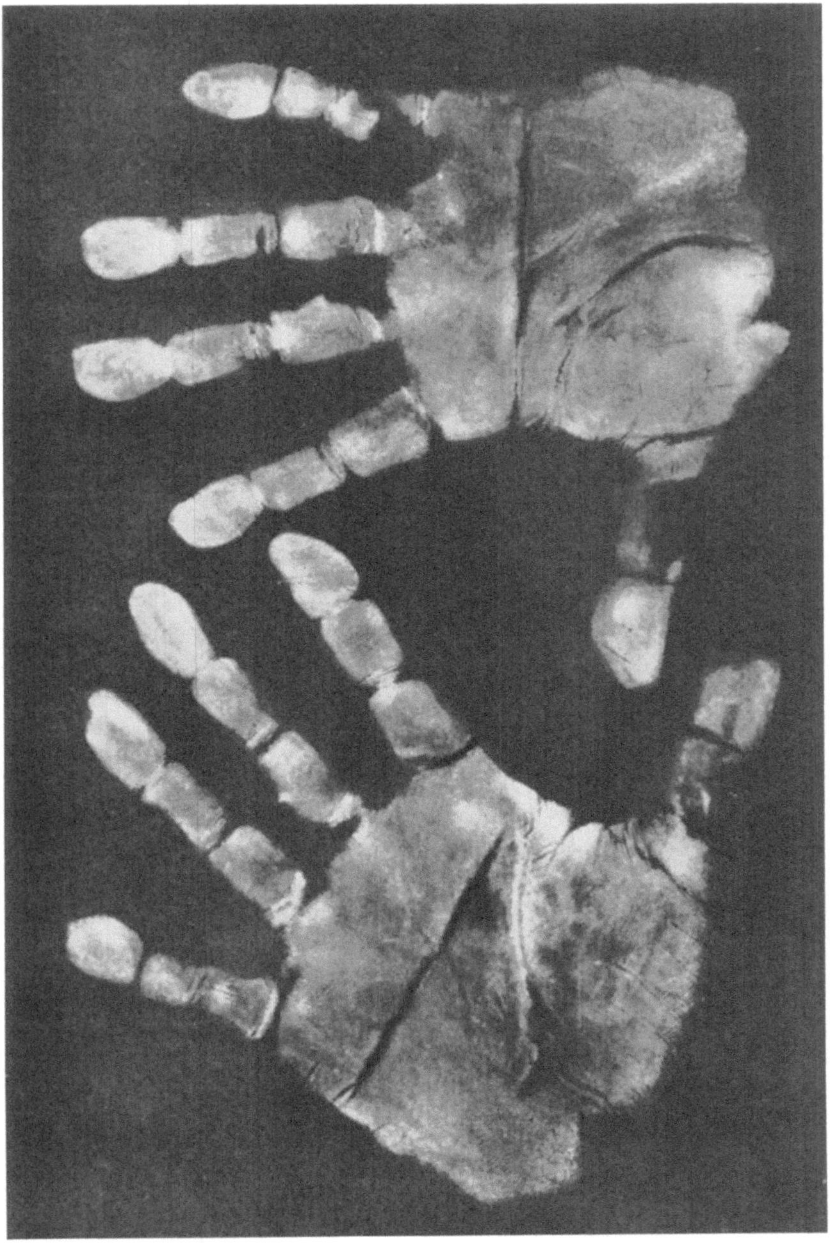

For technical reasons not all the prints are here produced in the original size.—The Editor.

PLATE II

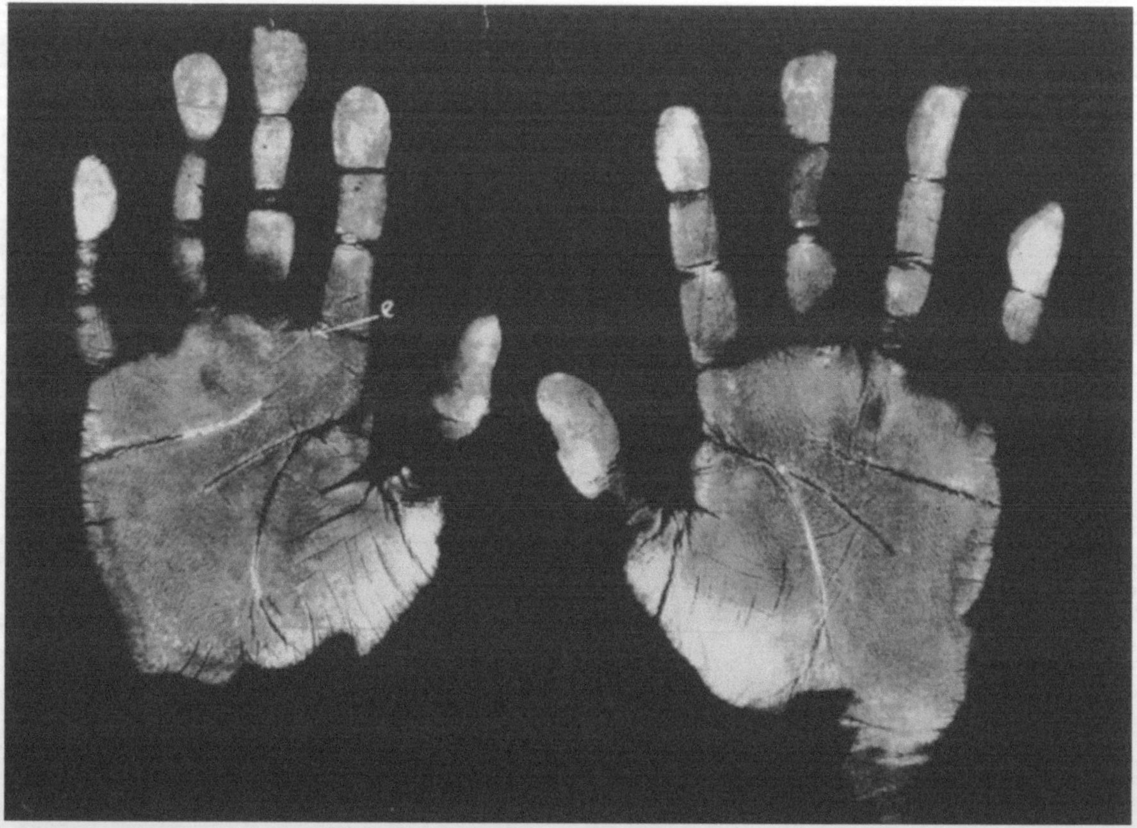

PLATE III

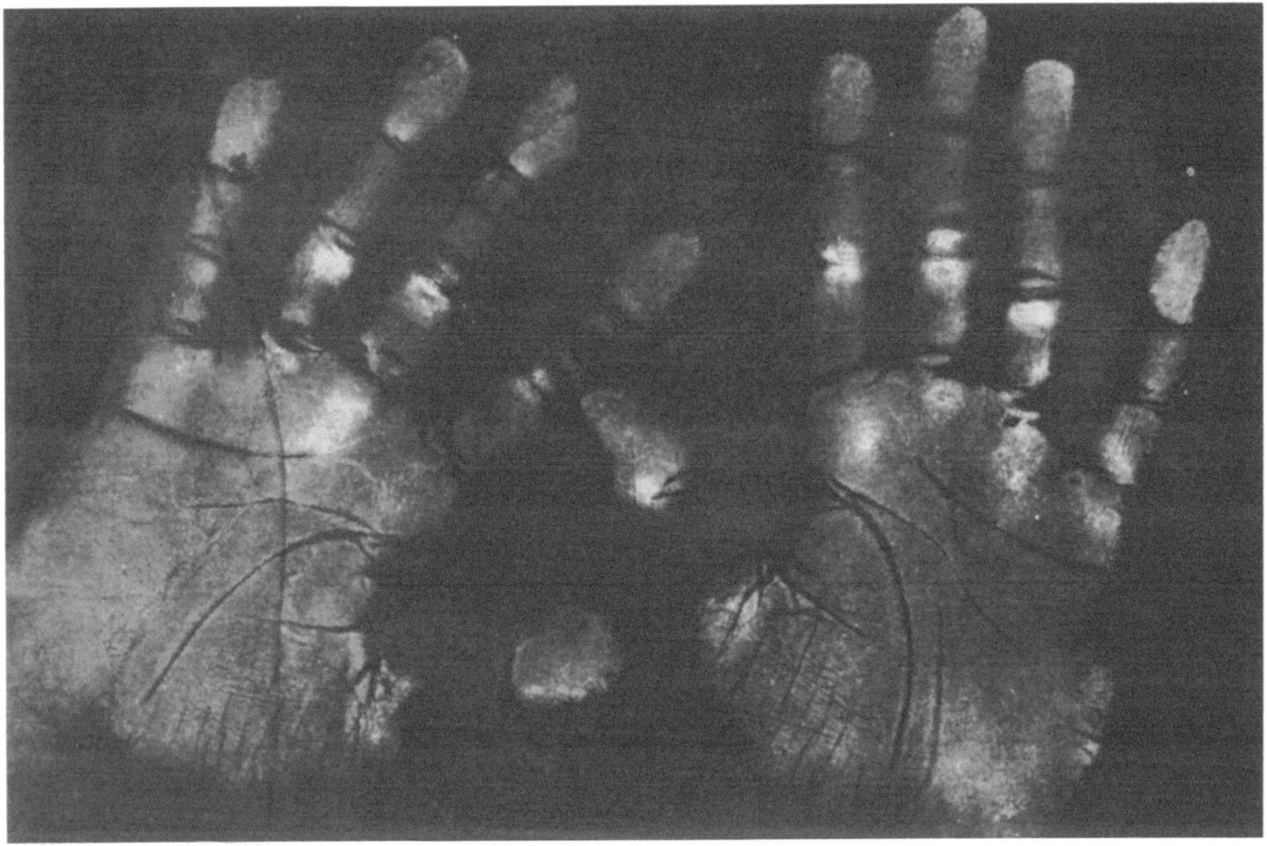

PLATE IV

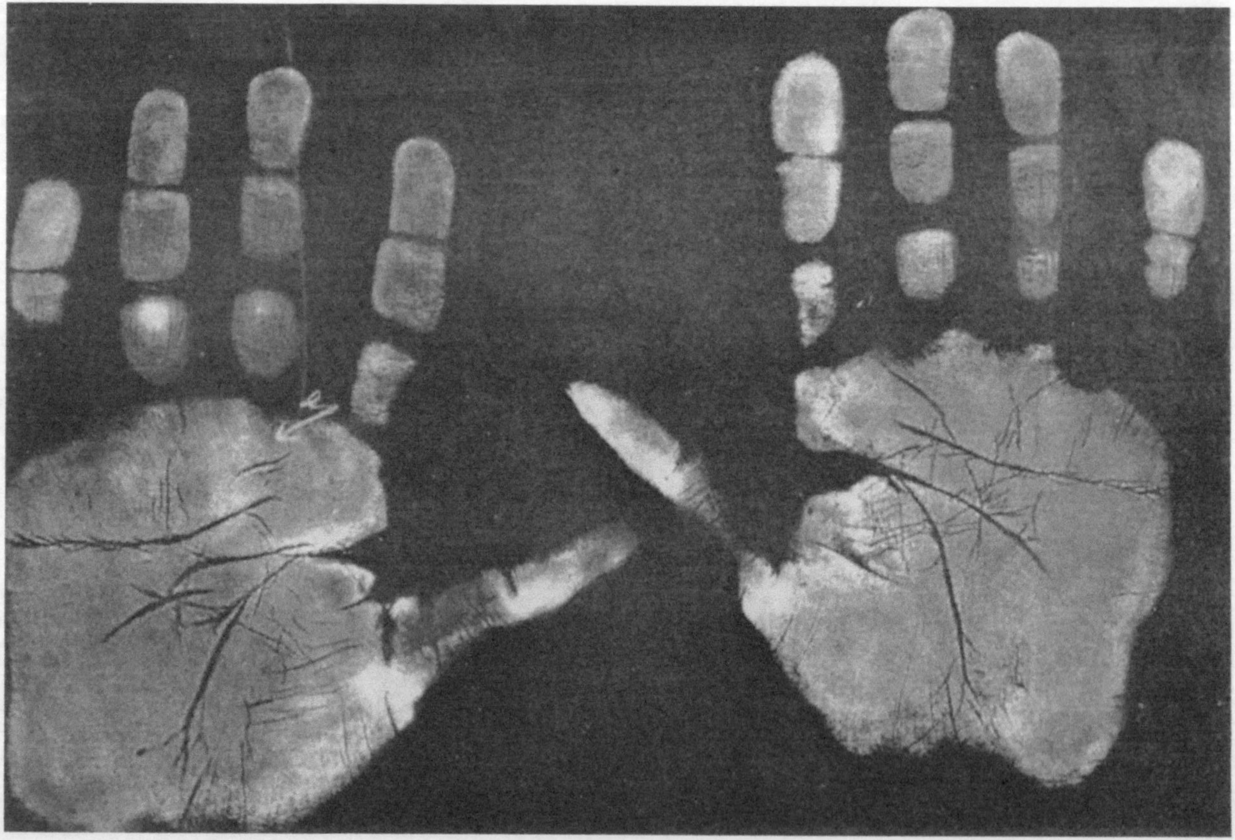

Plate V

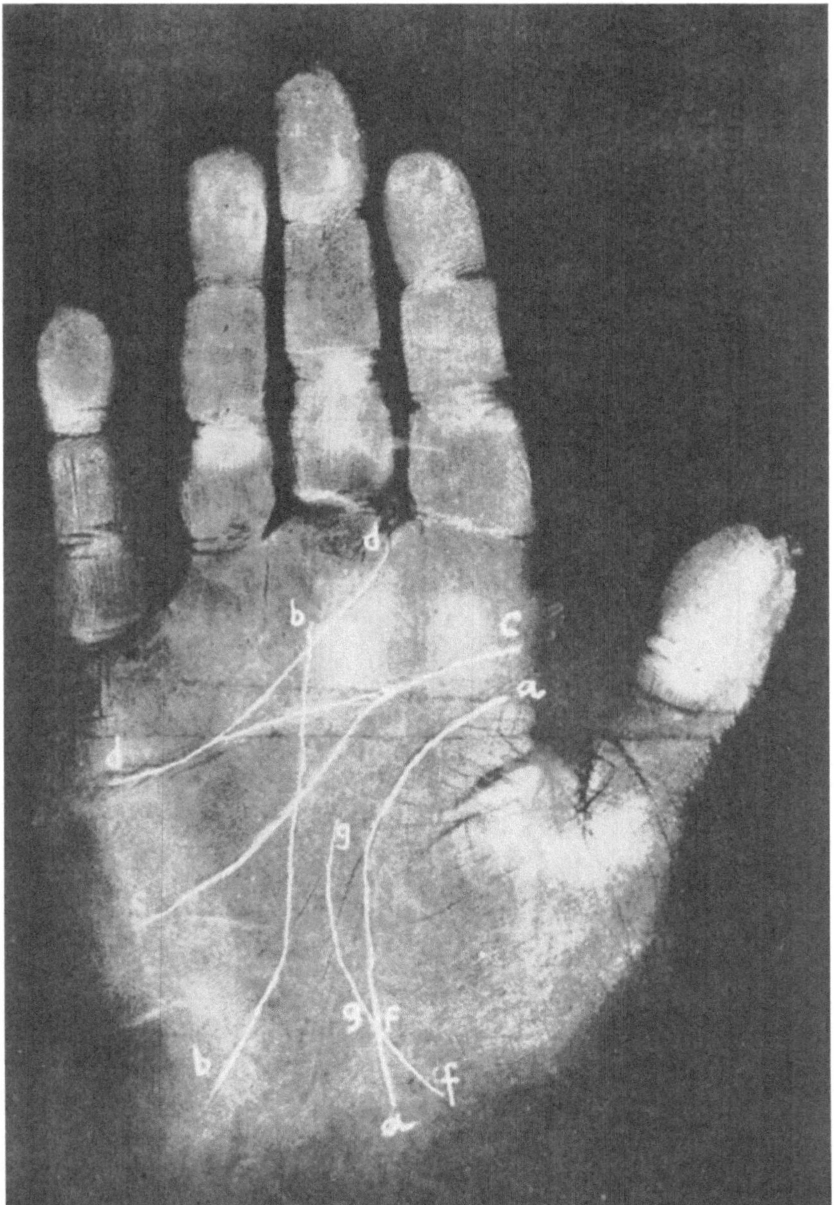

PLATE VI

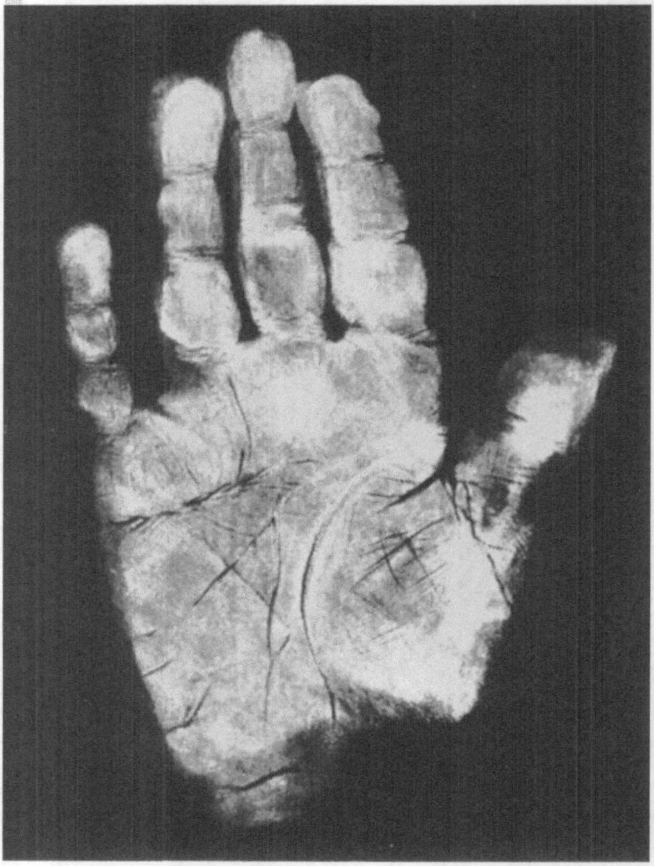

Plate VII

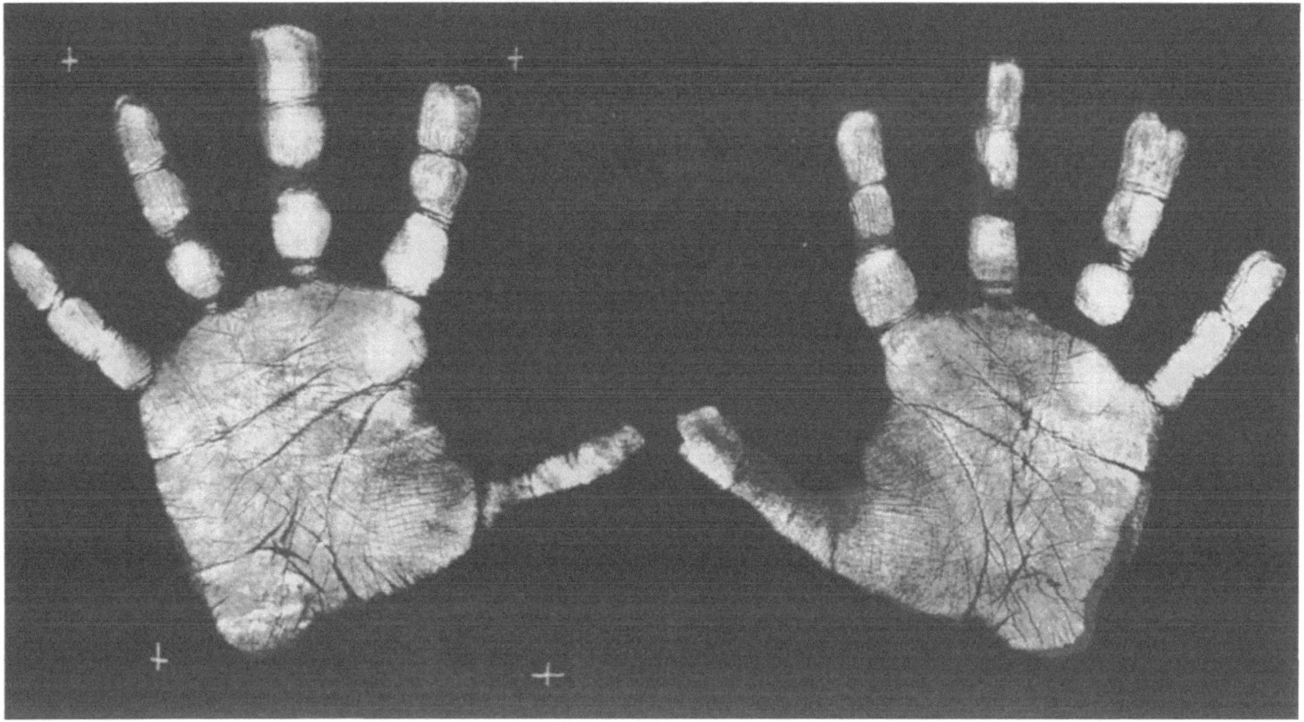

Plate VIII

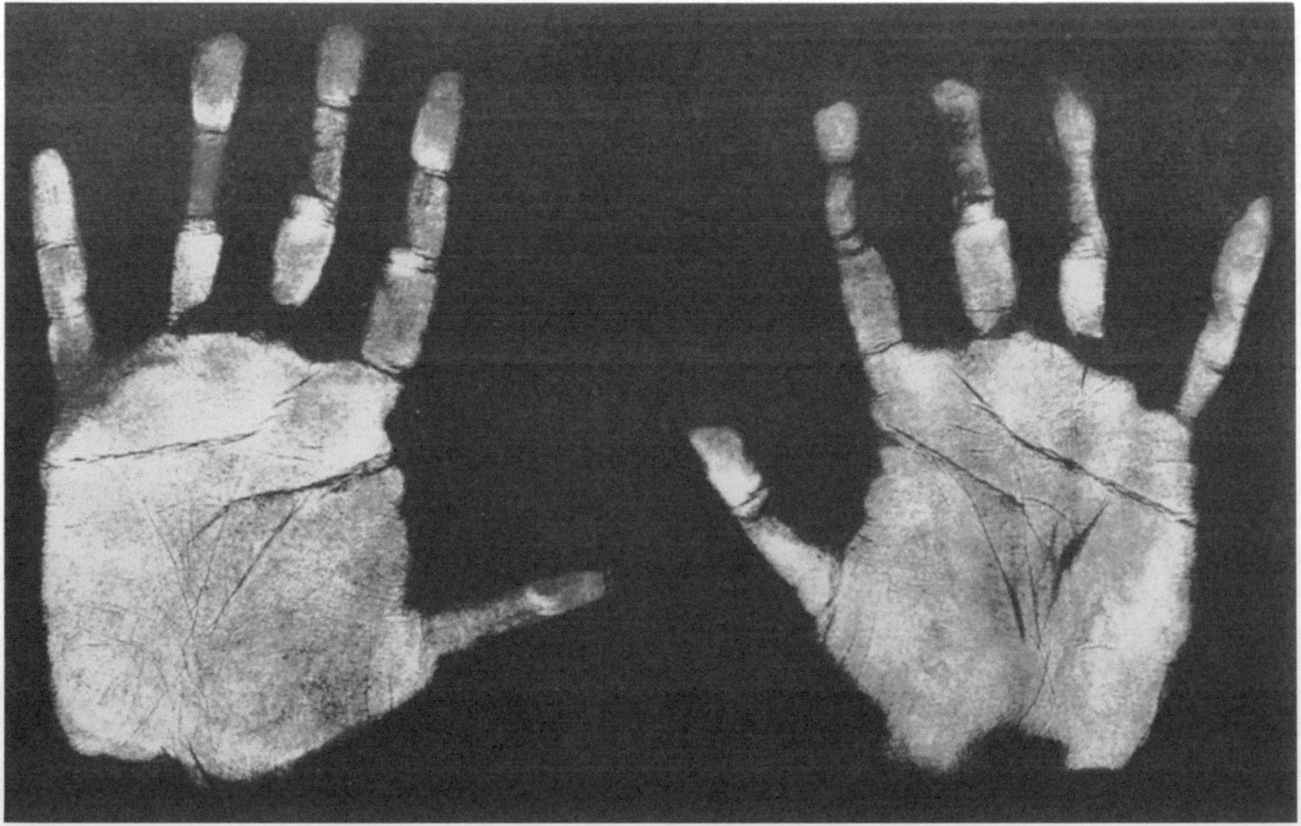

Plate IX

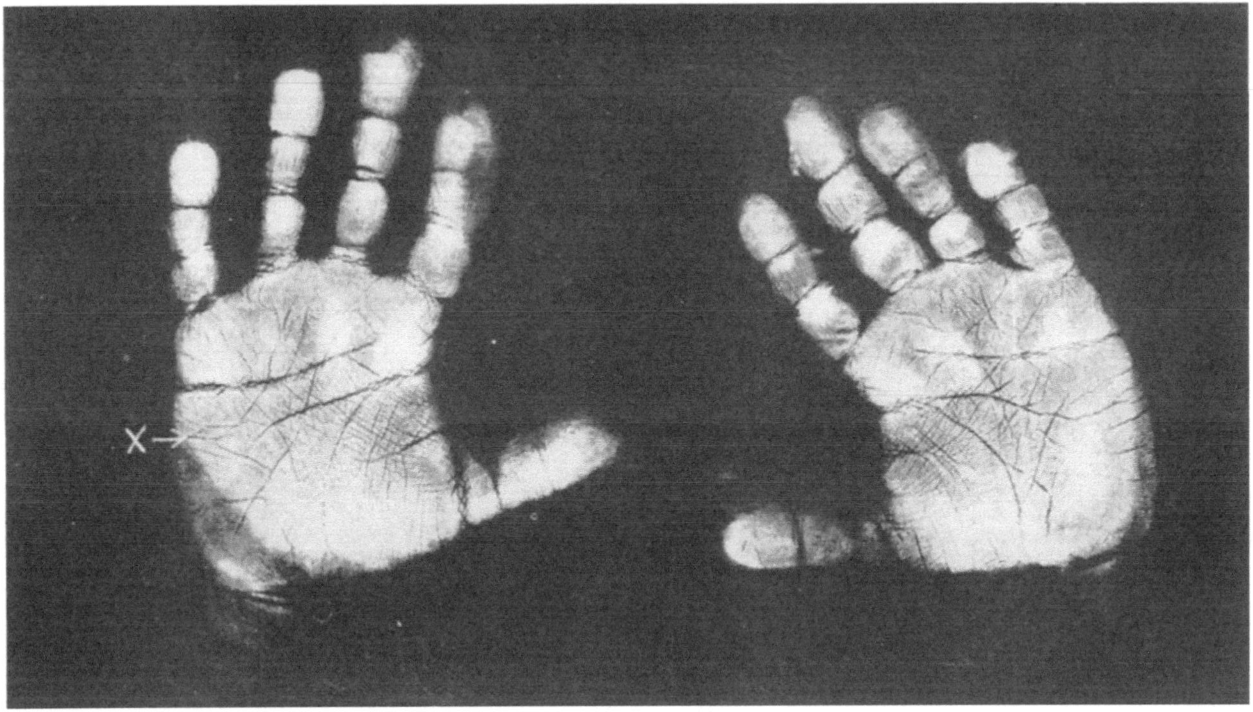

Plate X

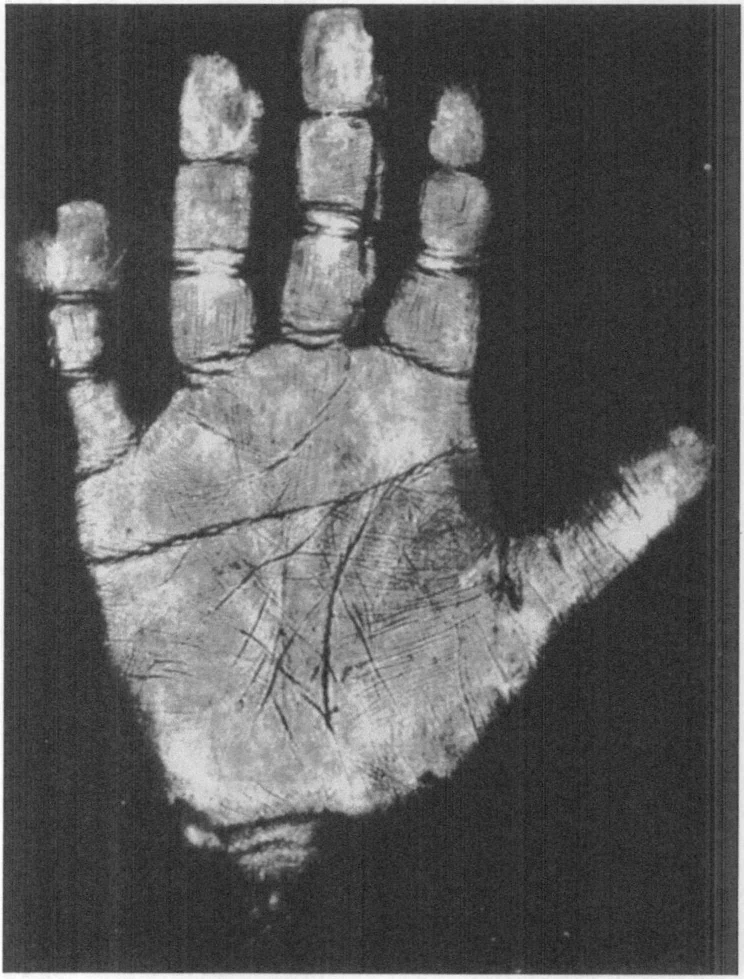

PLATE XI

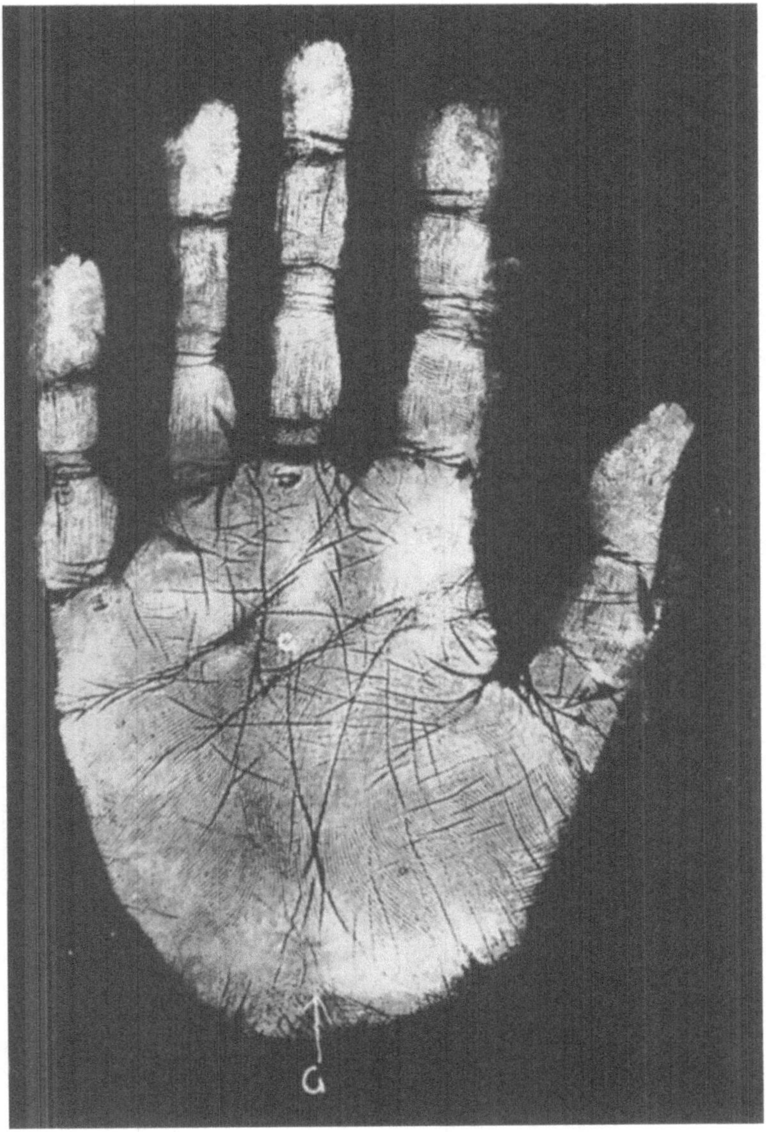

PLATE XII

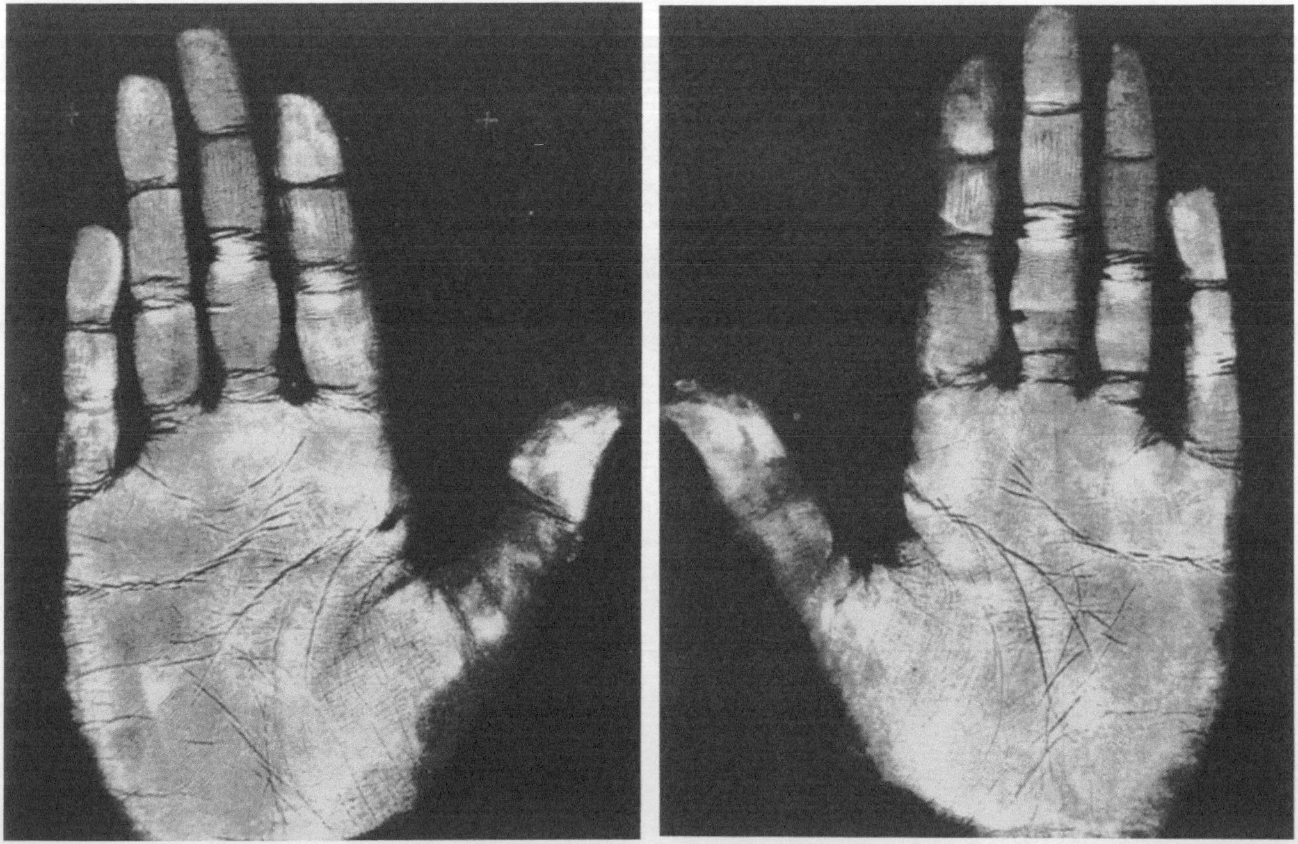

Plate XIII

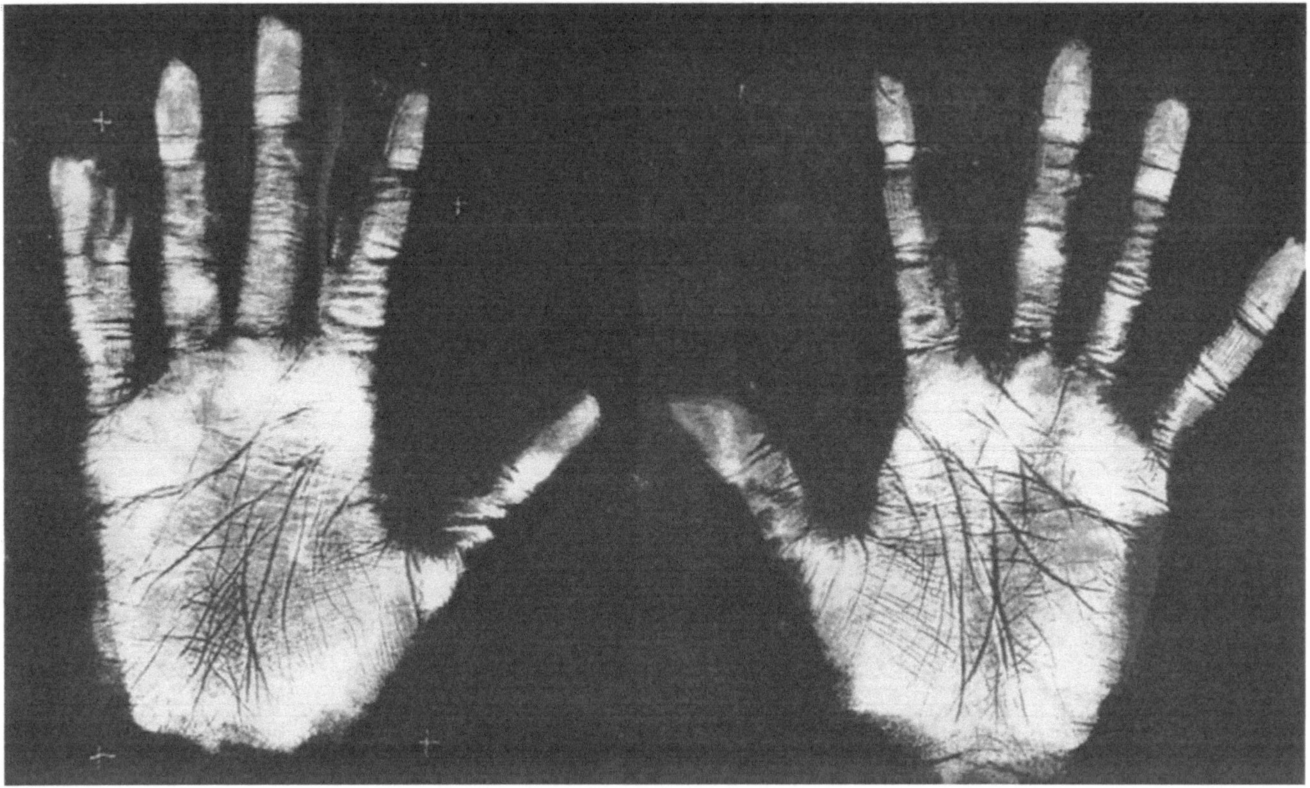

PLATE XIV PLATE XV PLATE XVI

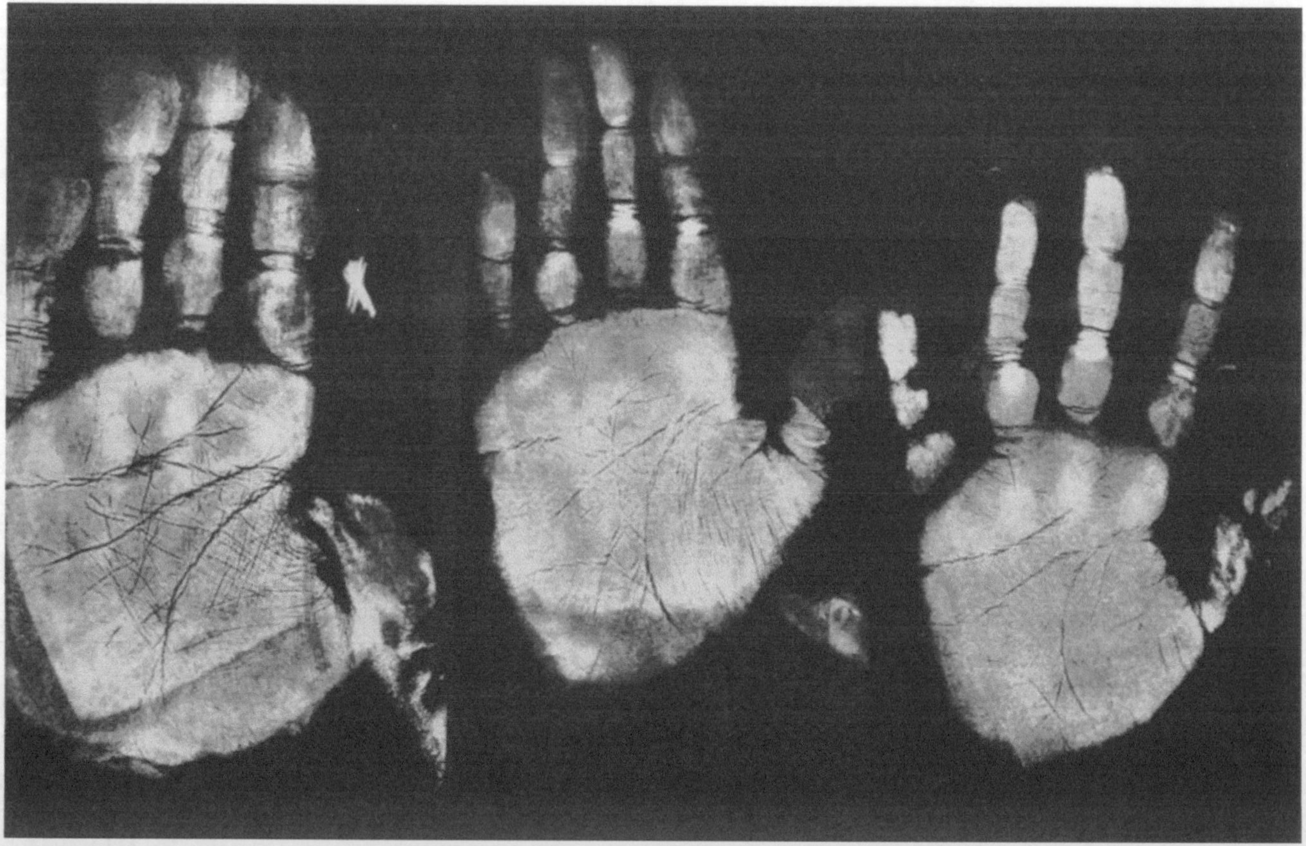

Plate XVII

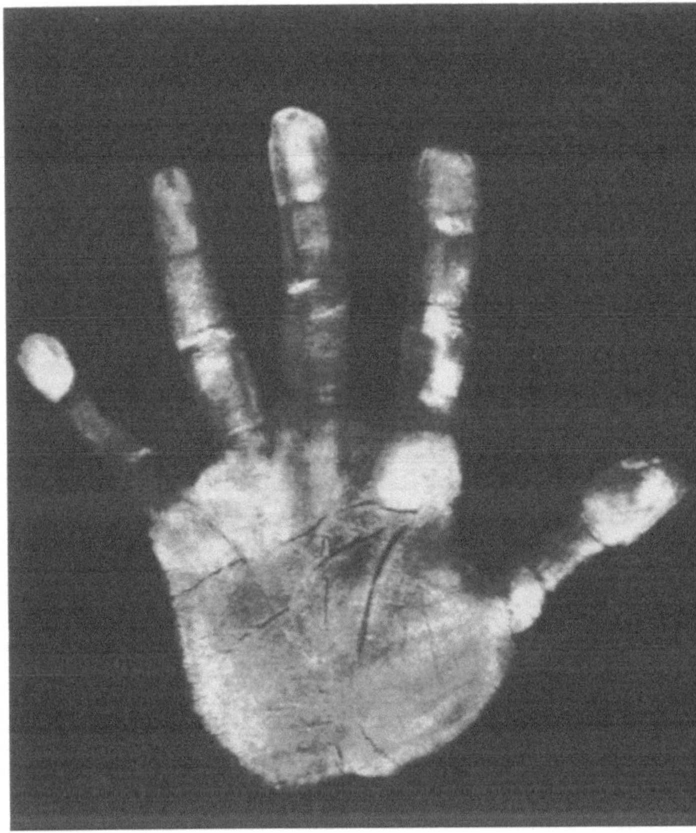

Plate XVIII

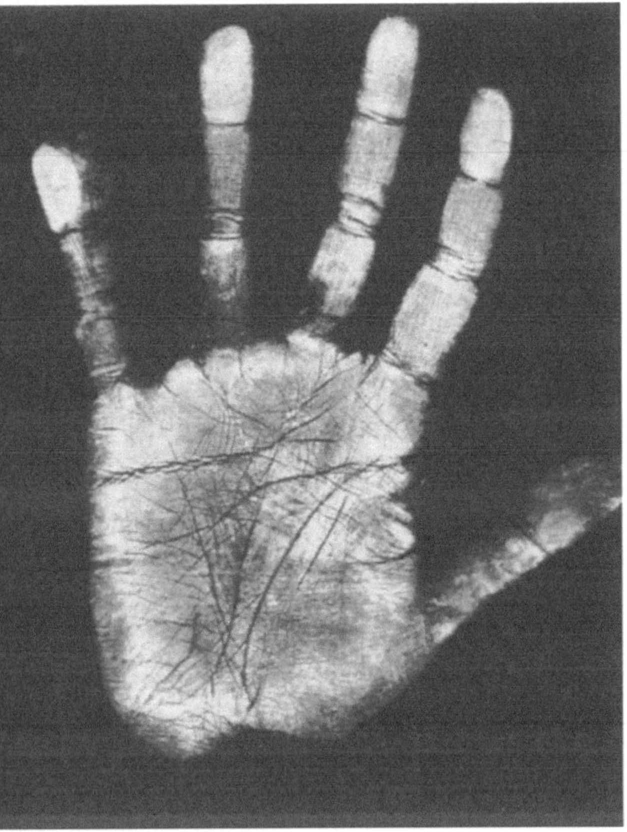

PLATE XIX

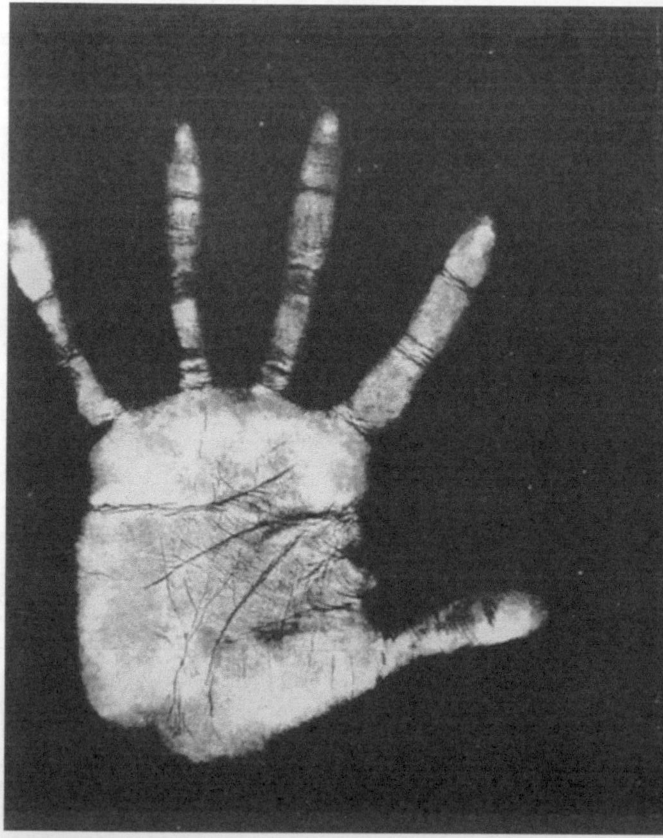

PLATE XX

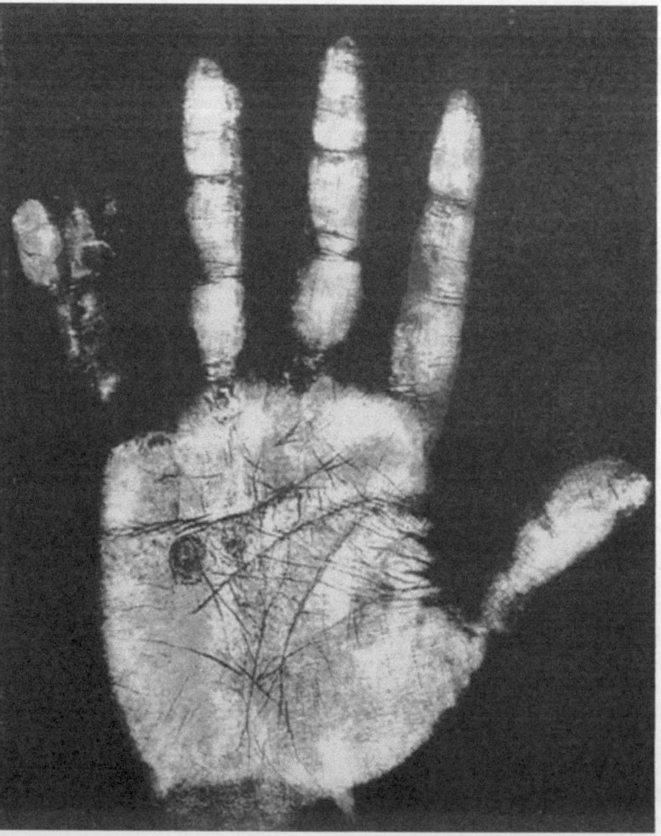

Plate XXI

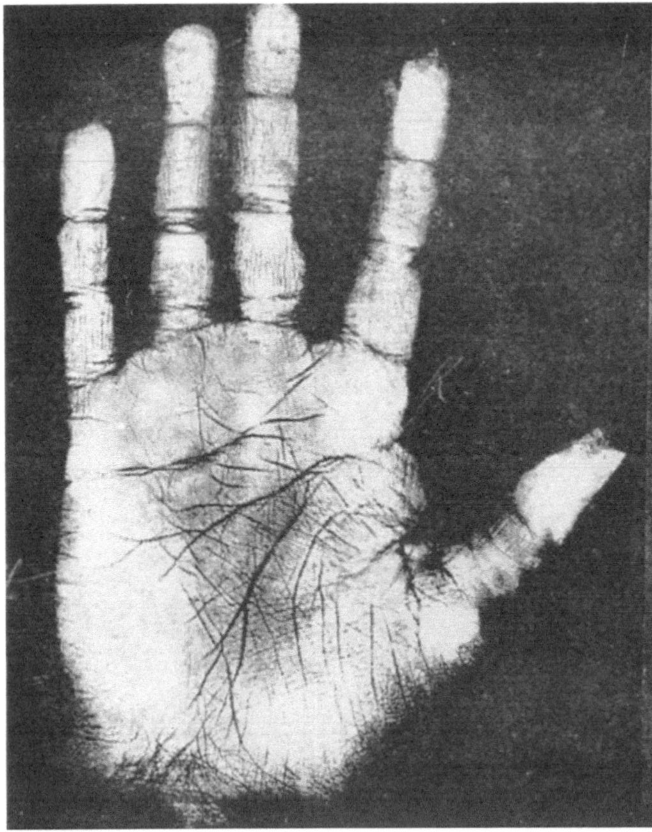

Plate XXII

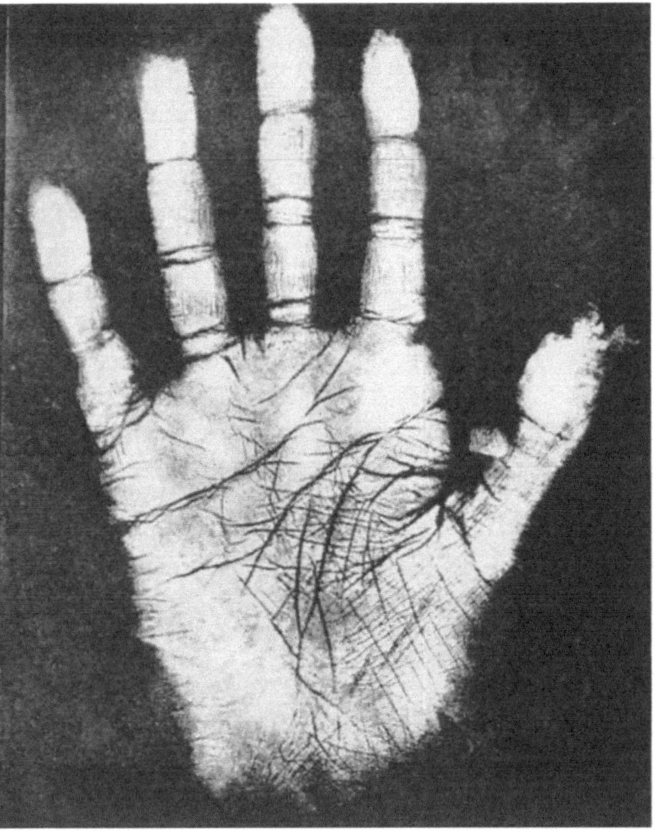

PLATE XXIII

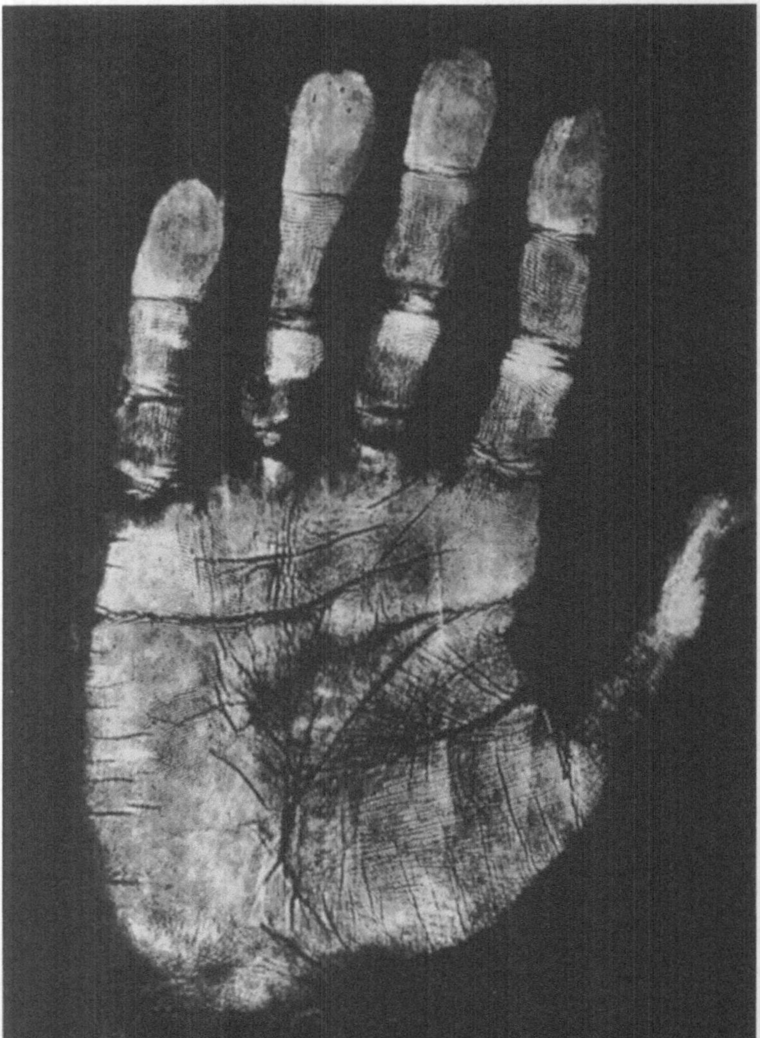

Plate XXIV

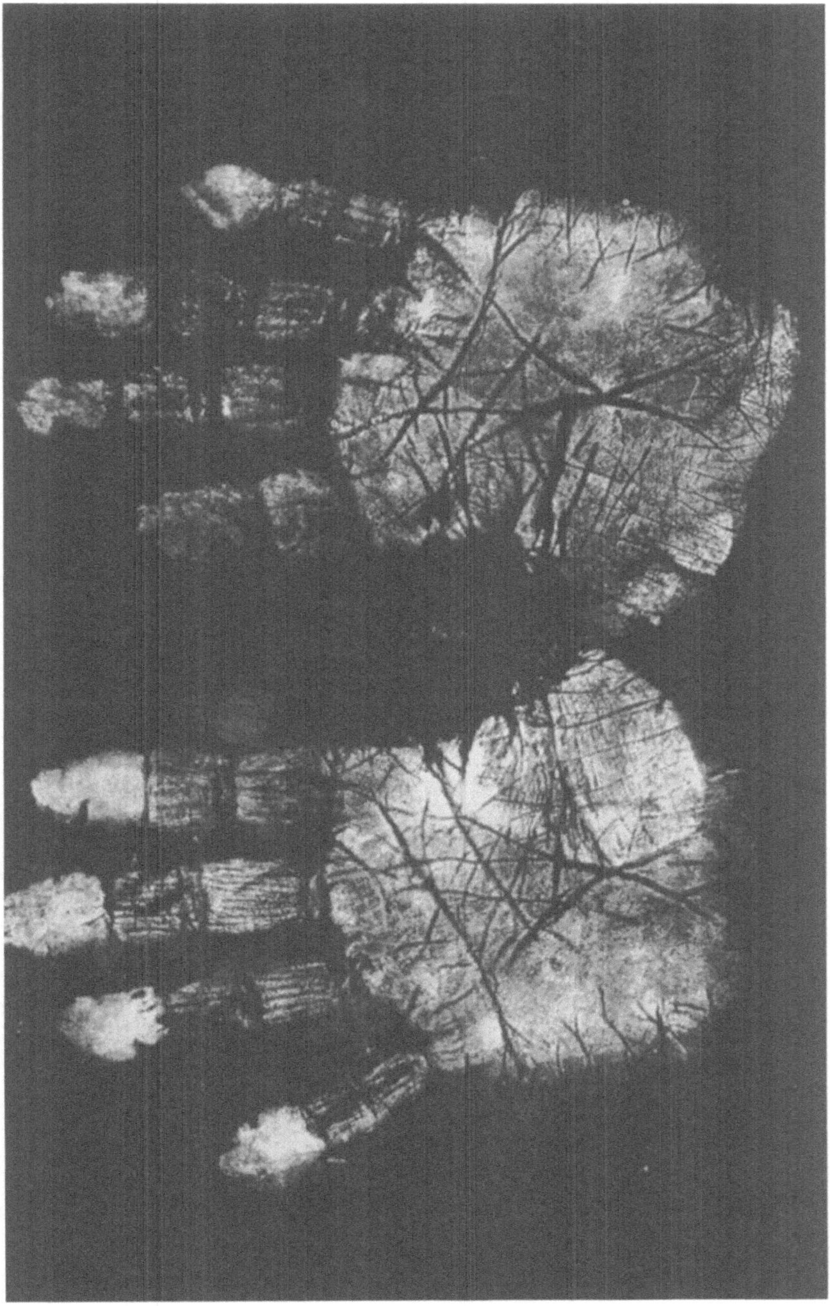

PLATE XXV

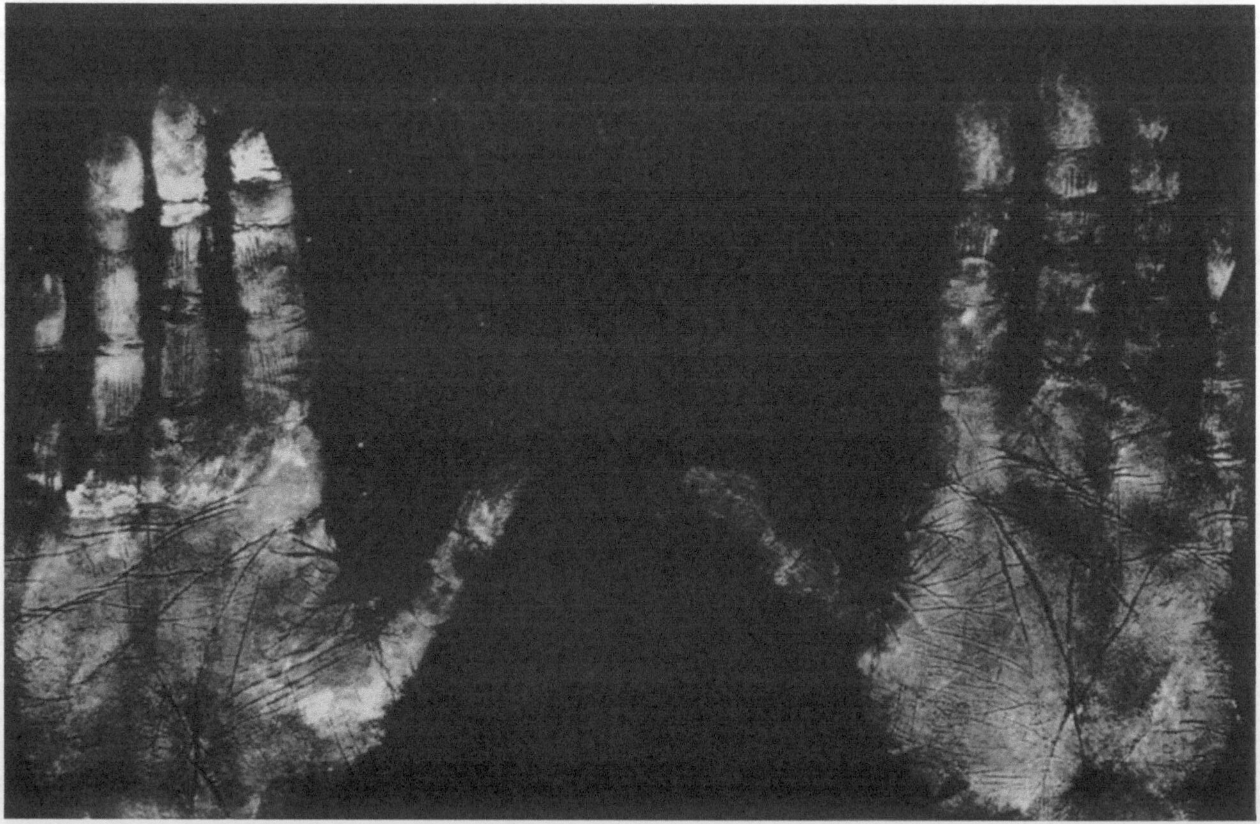

Plate XXVI

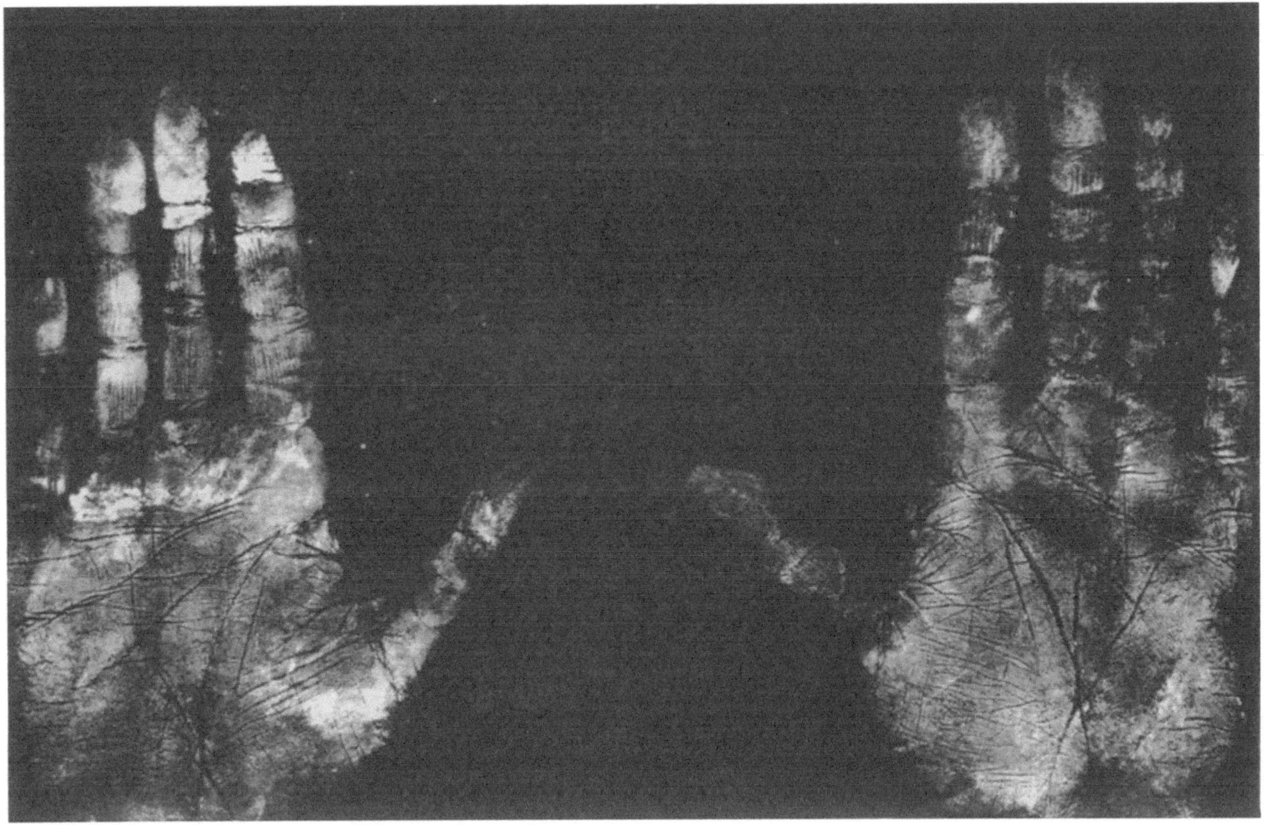

PLATE XXVII

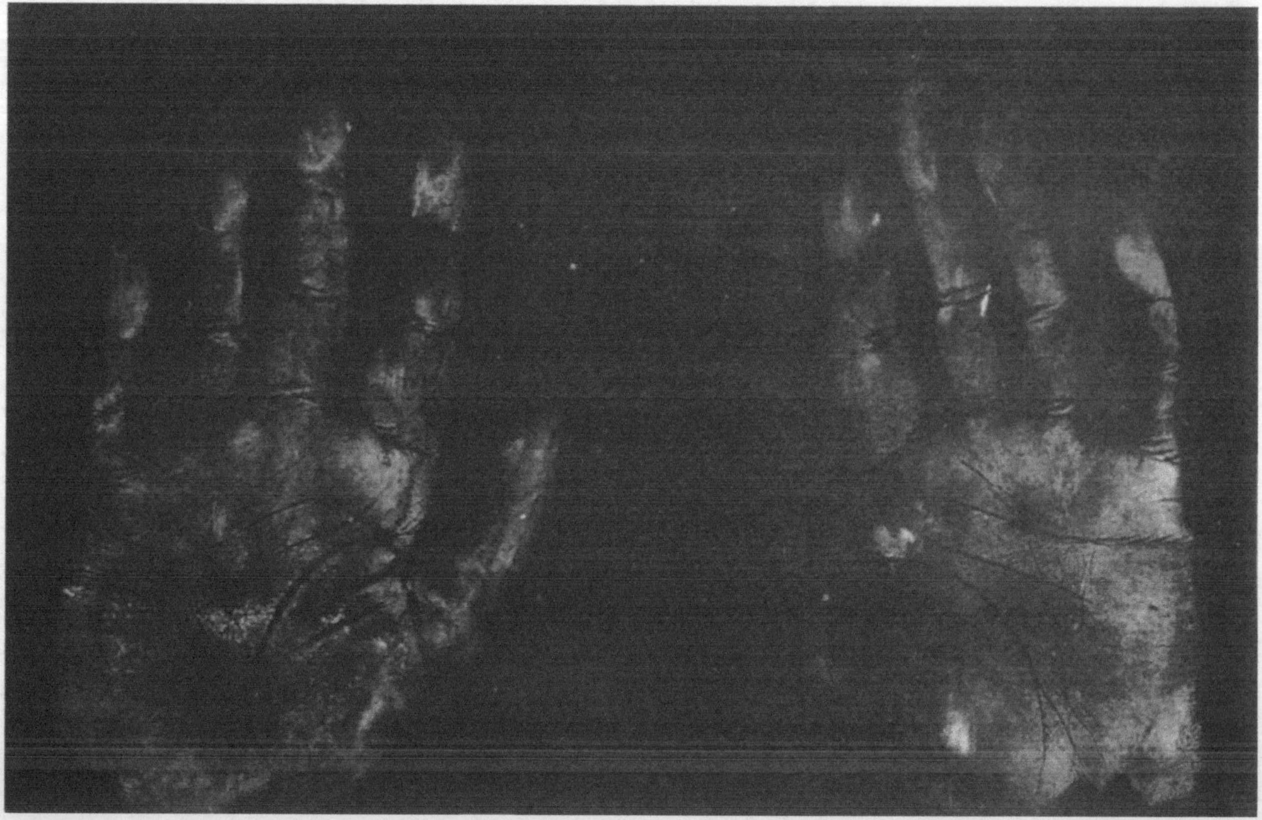

PLATE XXVIII

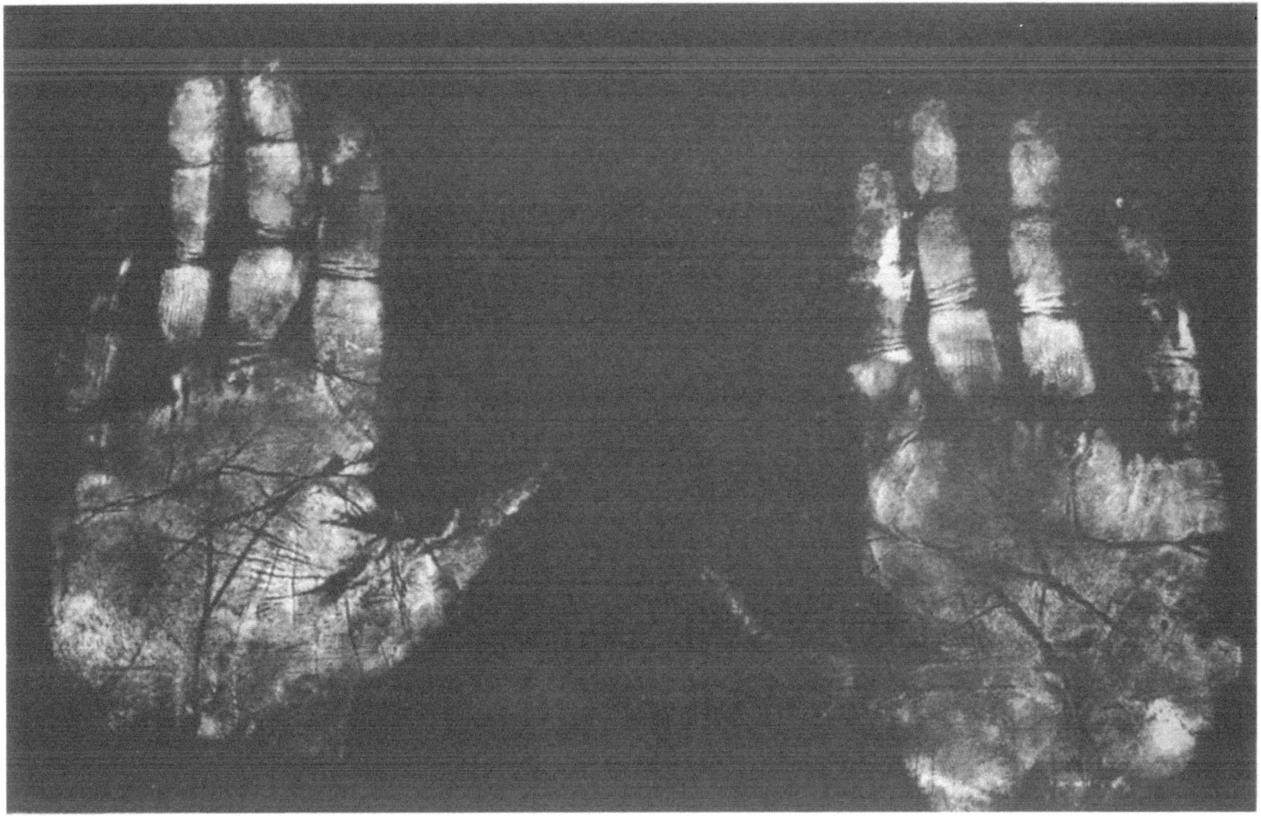

Plate XXIX

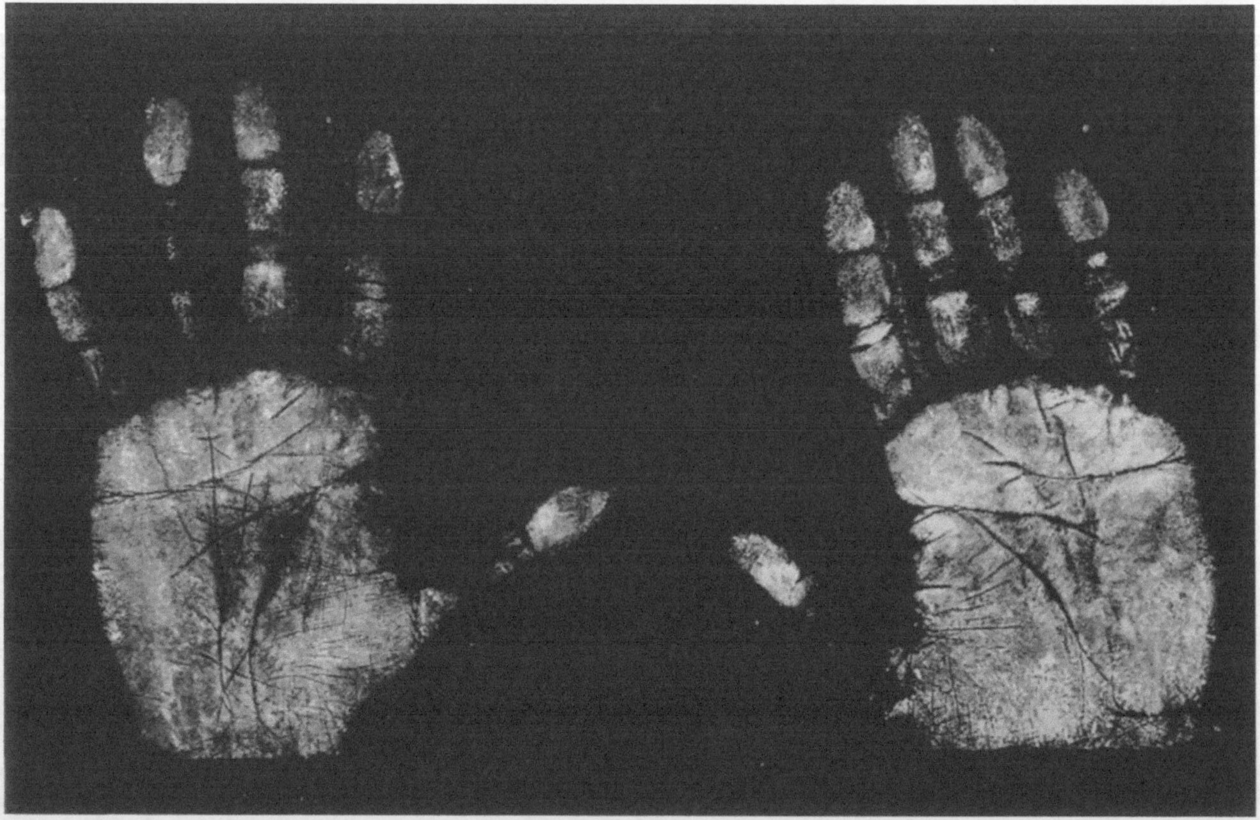

Plate XXX

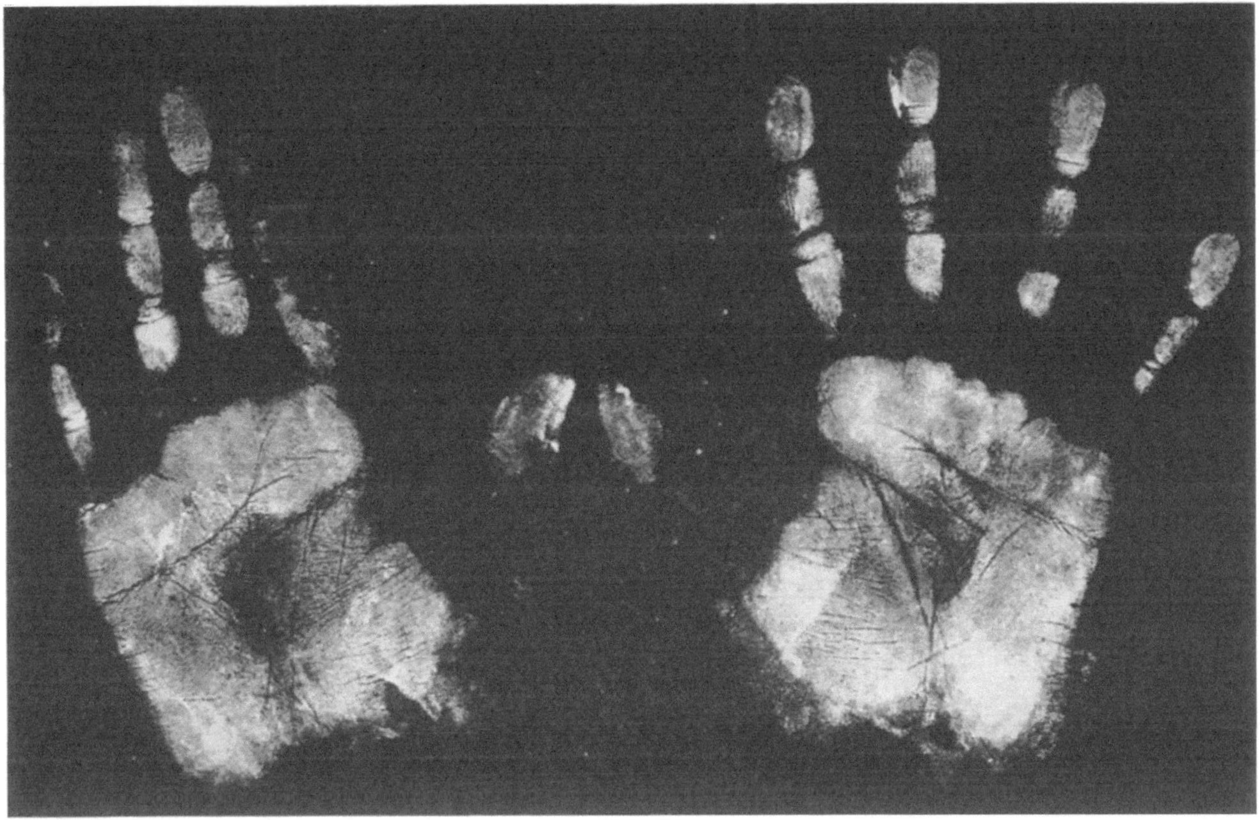

Plate XXXI

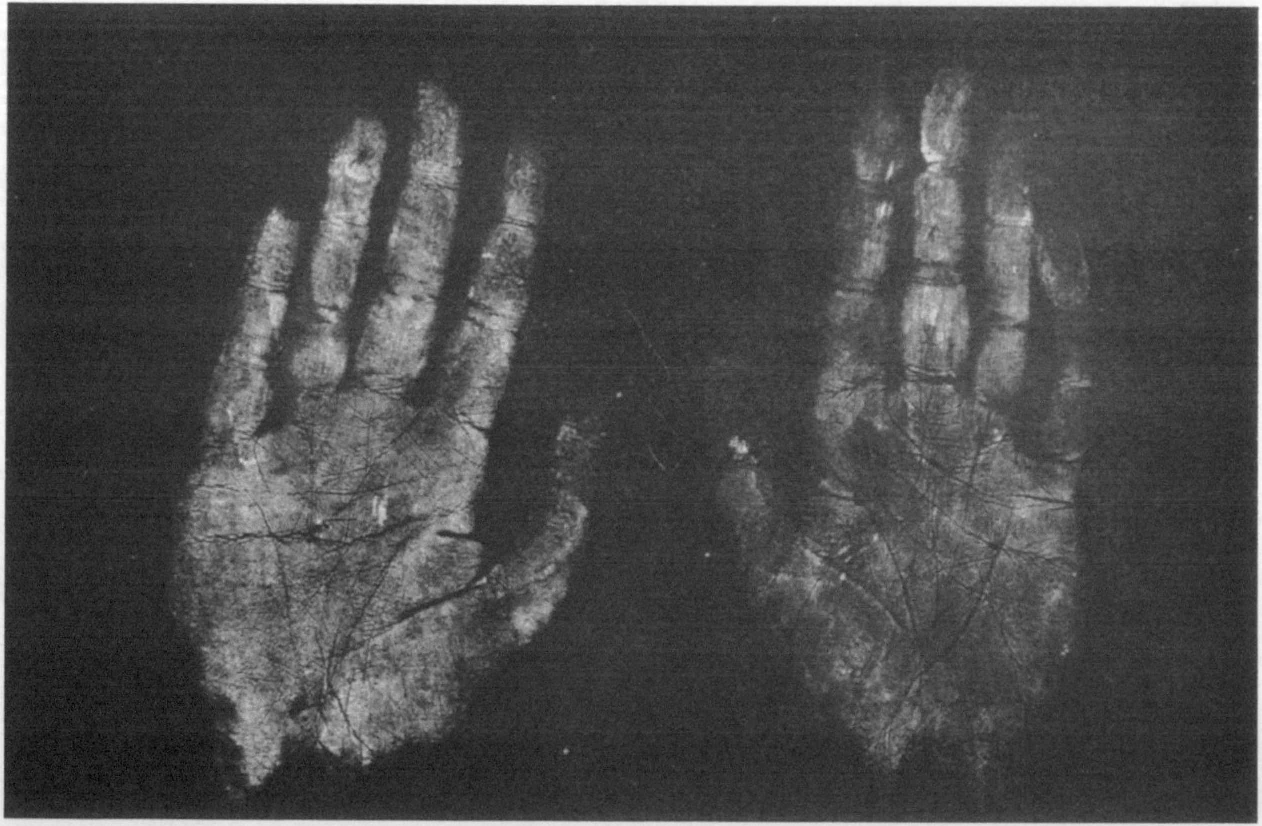

PLATE XXXII

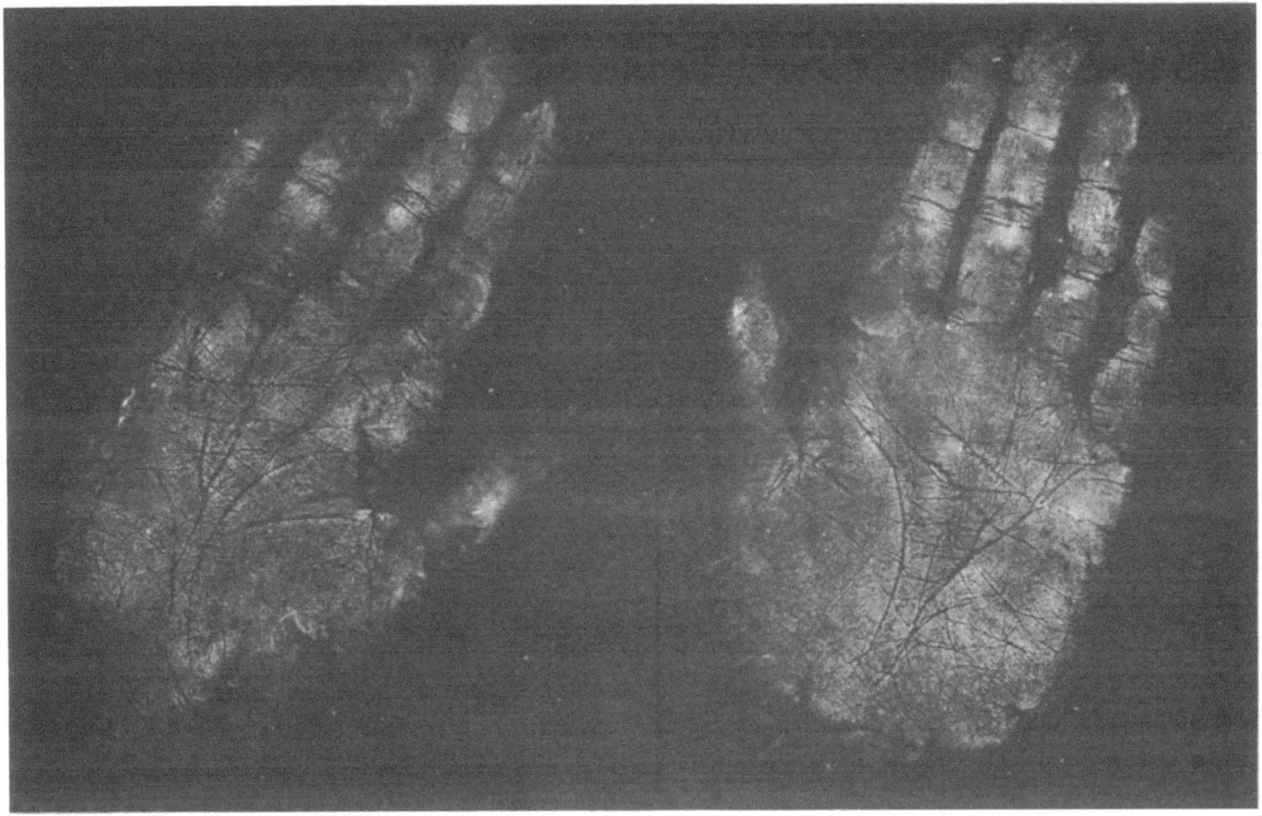

PLATE XXXIII

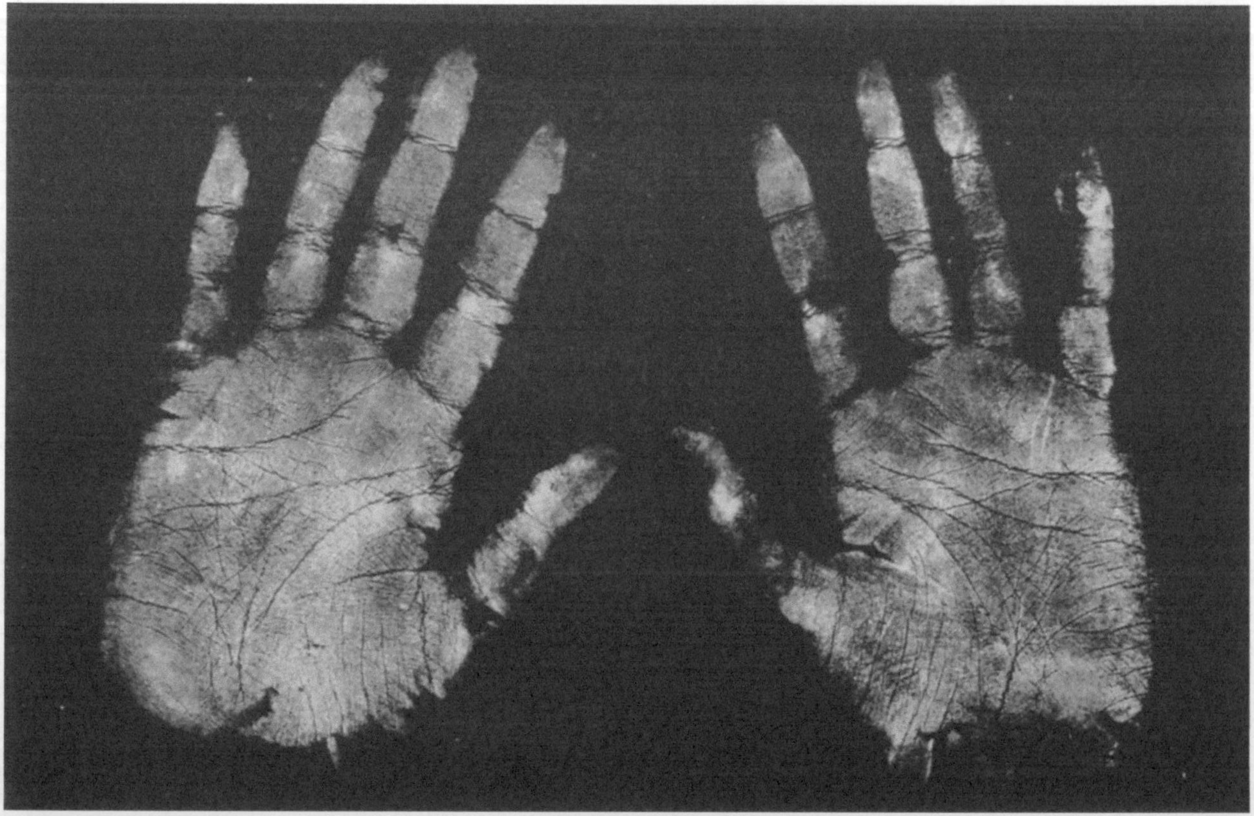

Plate XXXIV

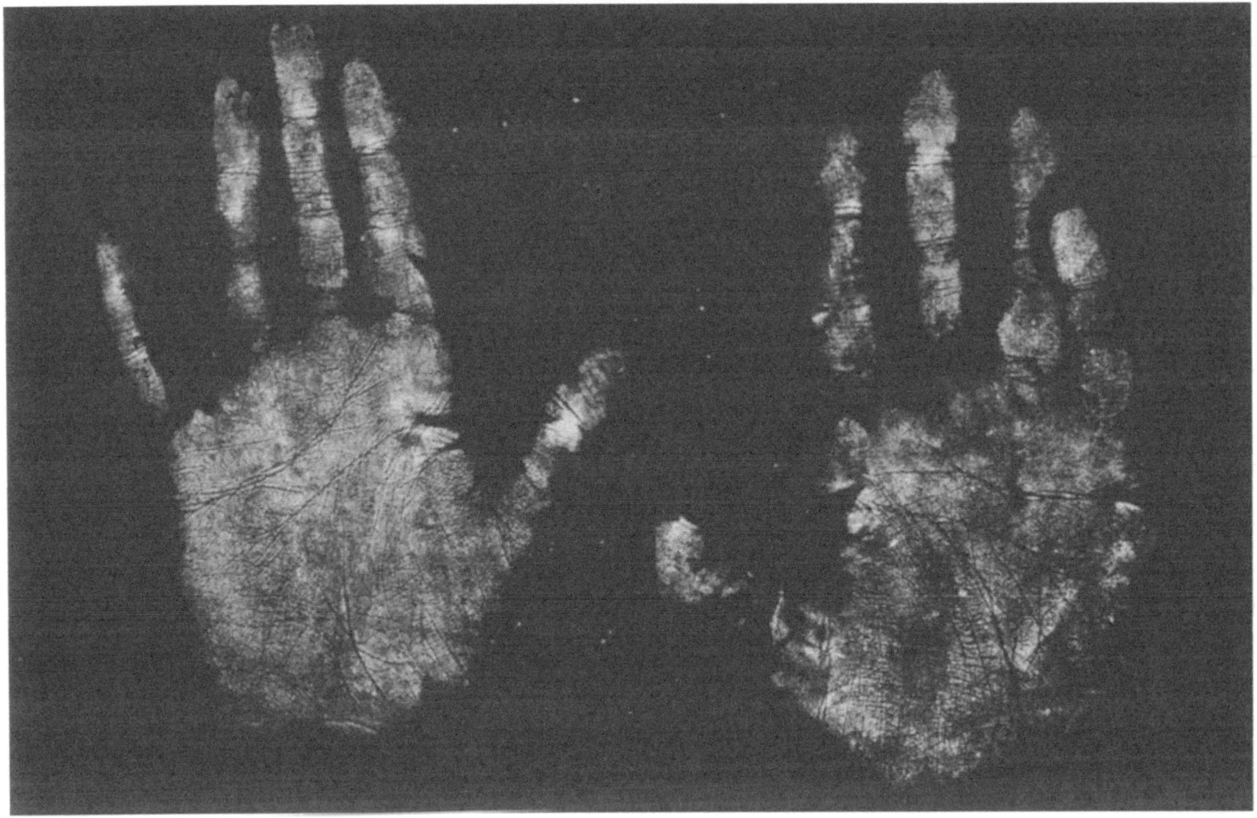

Plate XXXV

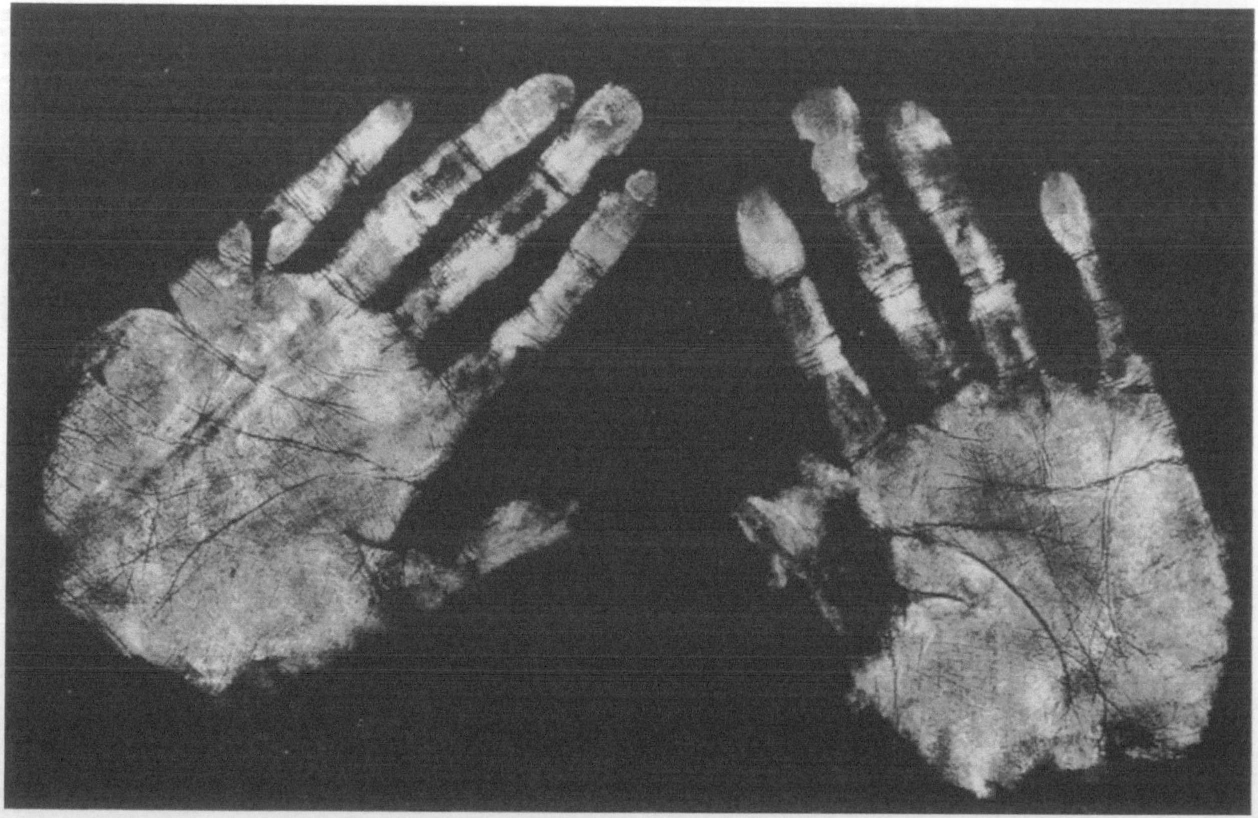

PLATE XXXVI

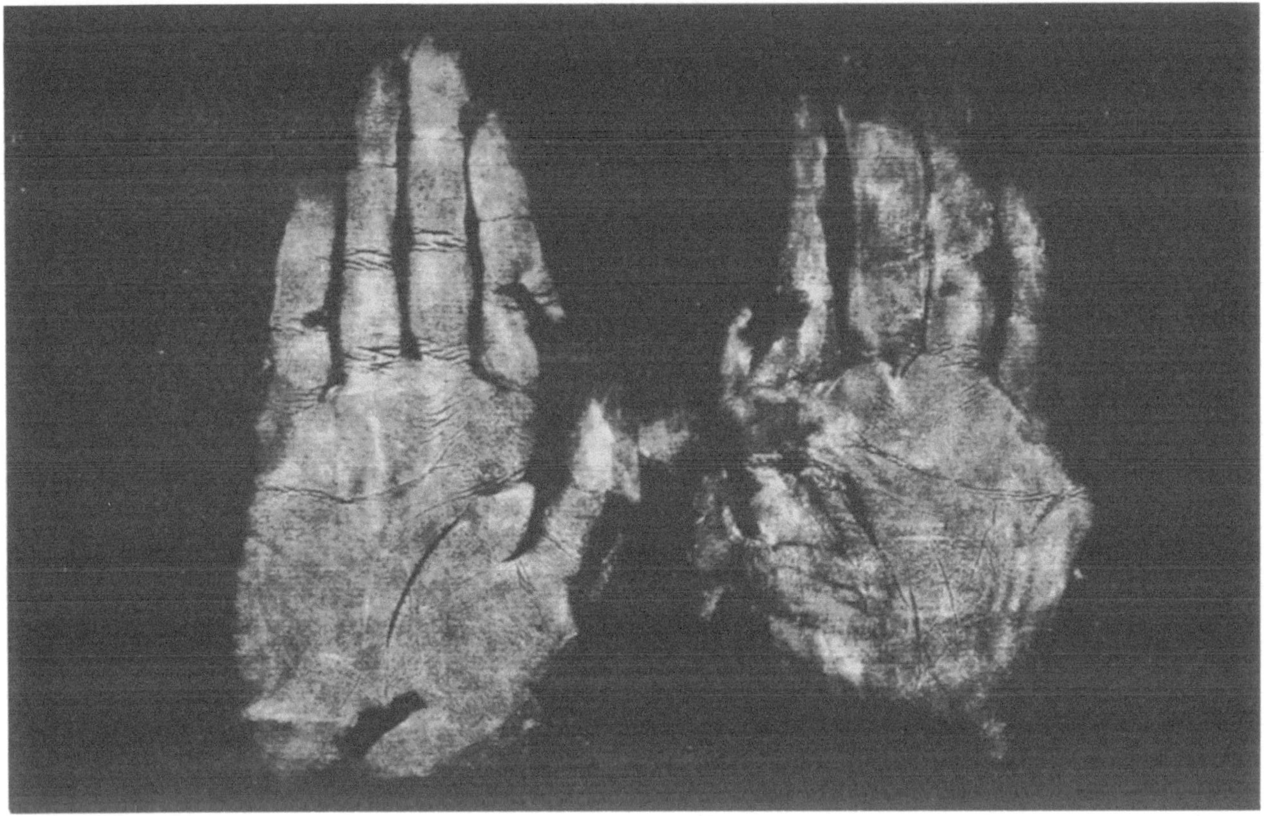

Plate XXXVII

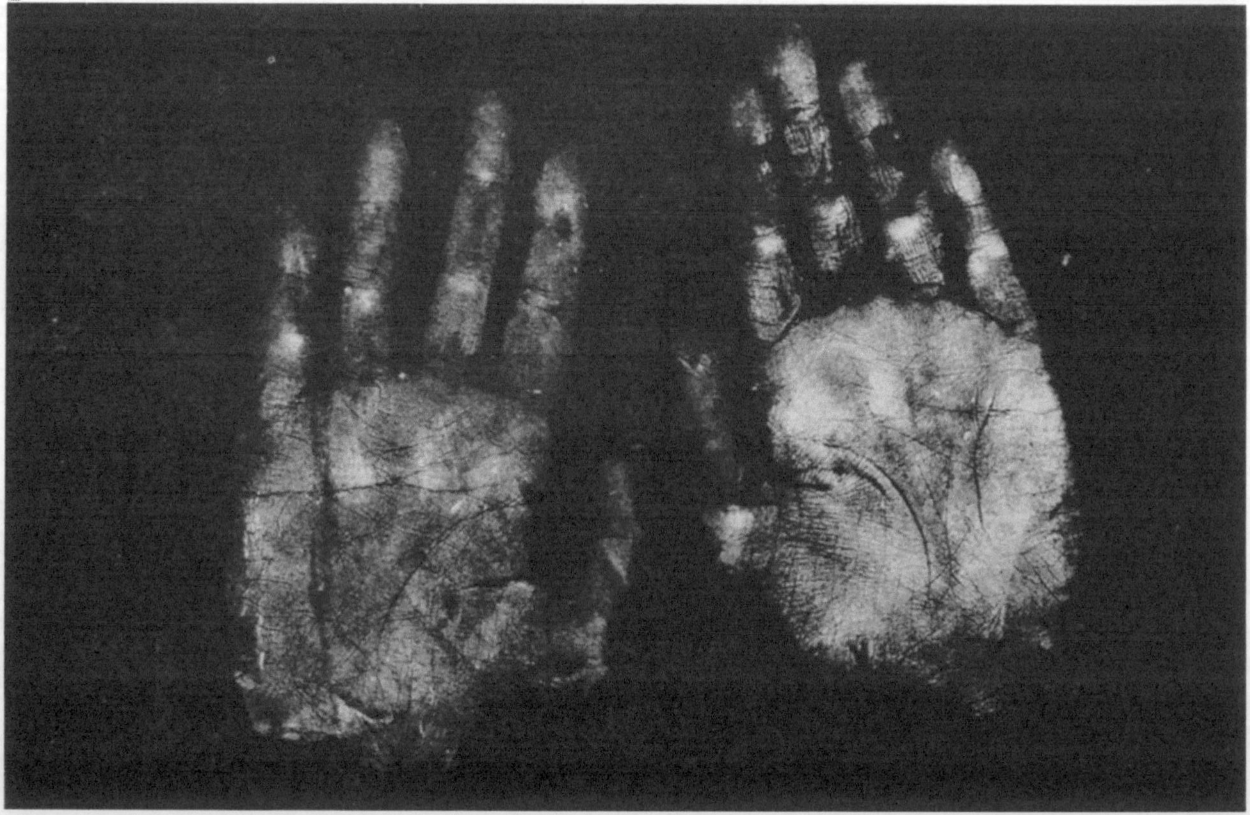

PLATE XXXVIII

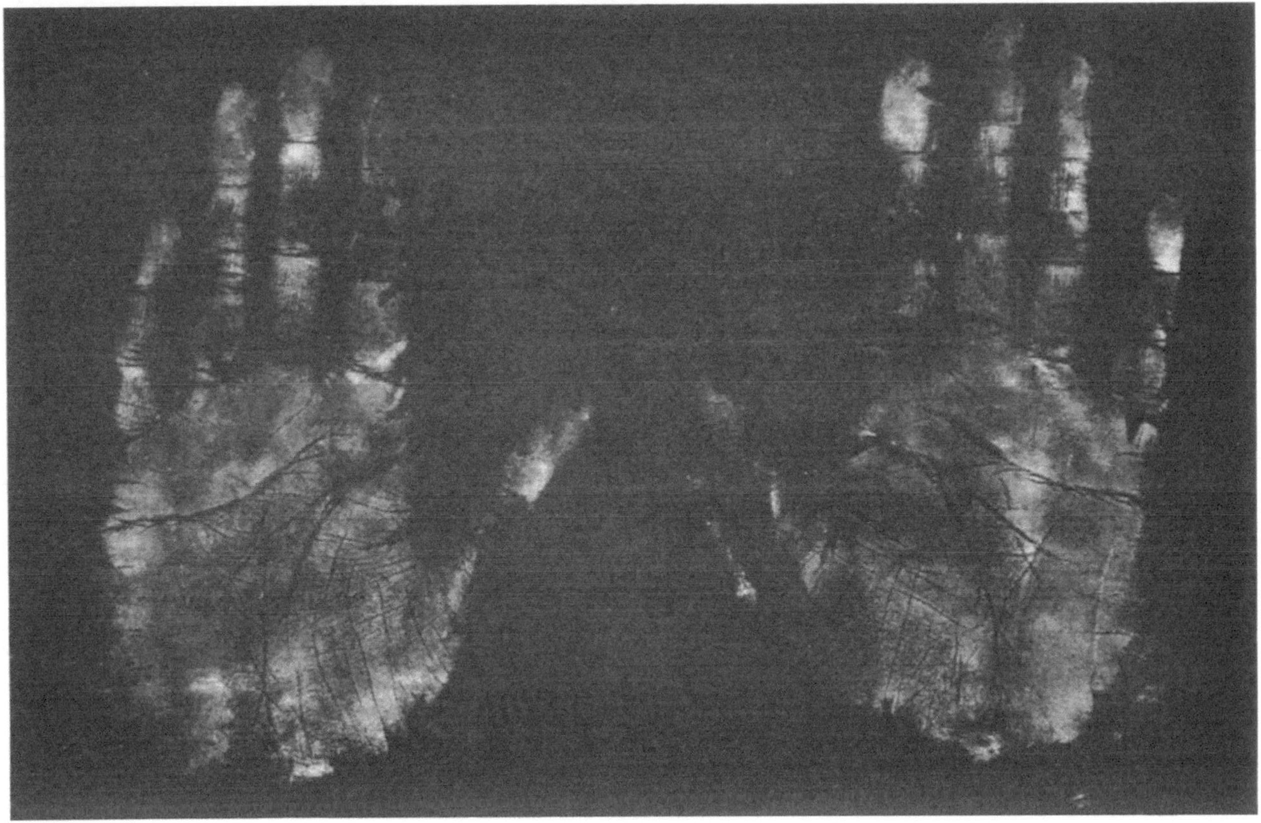

PLATE XXXIX

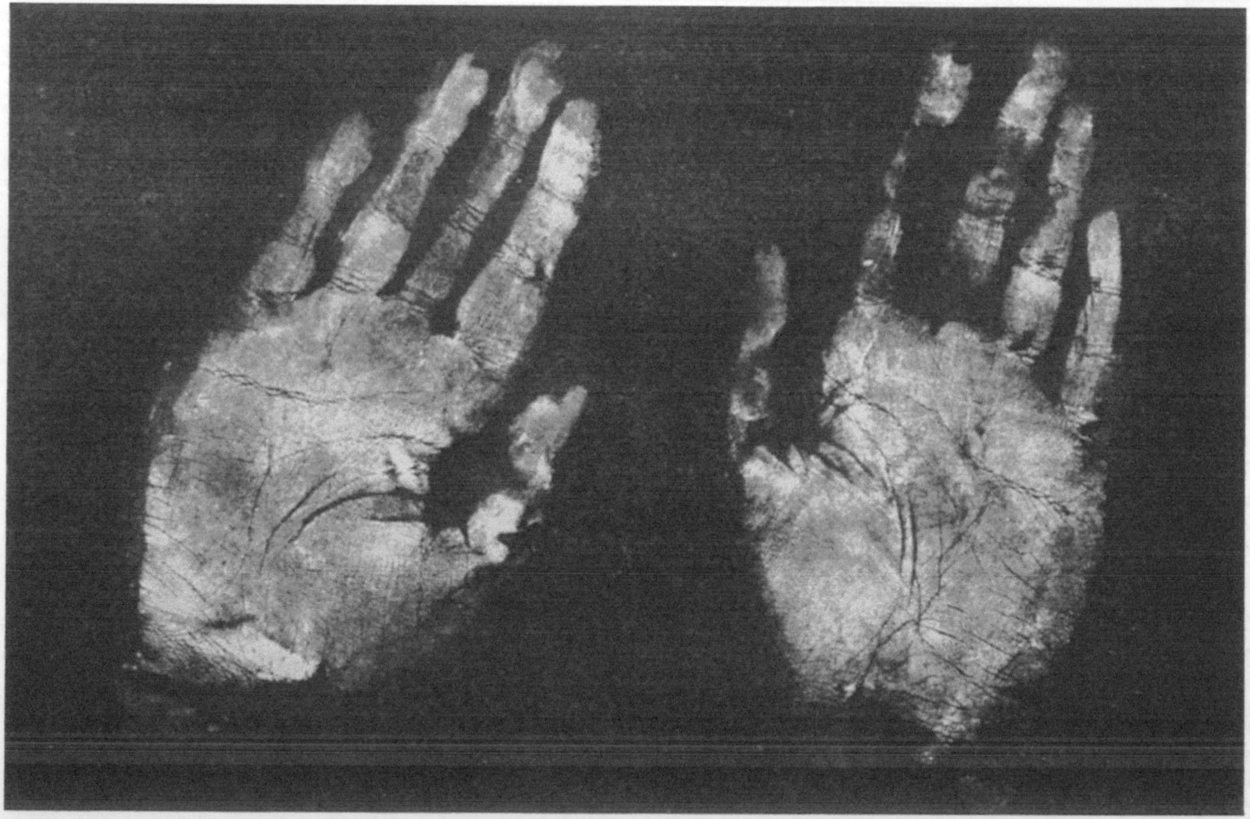

PLATE XL

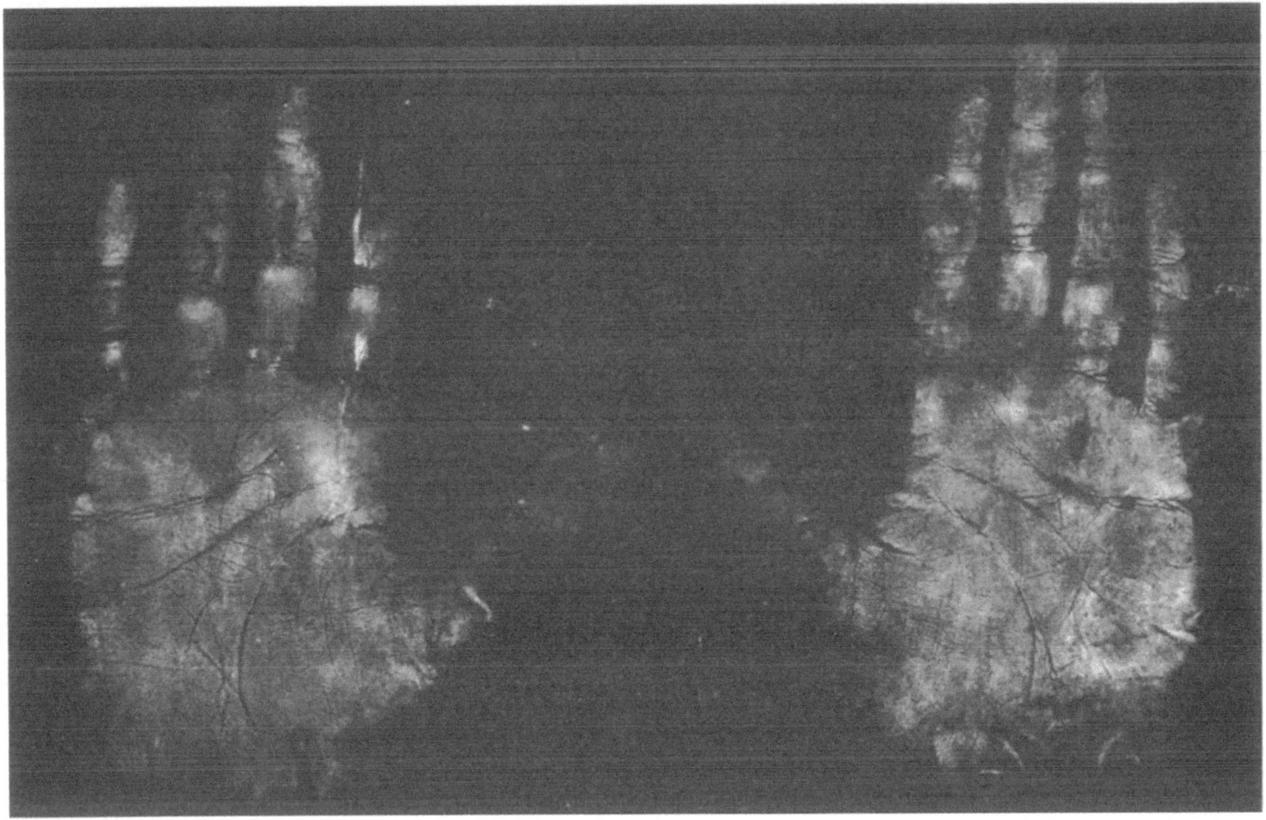

Plate XLI

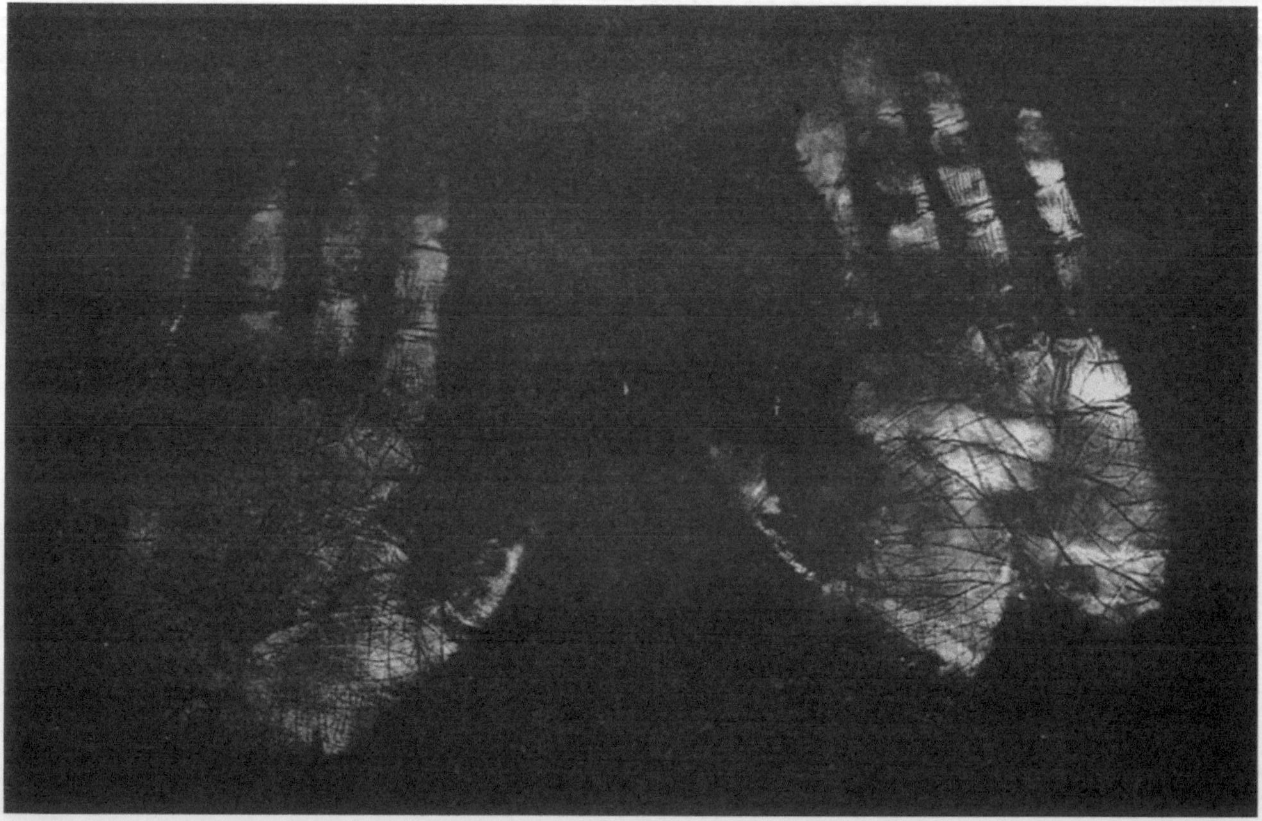

Plate XLII

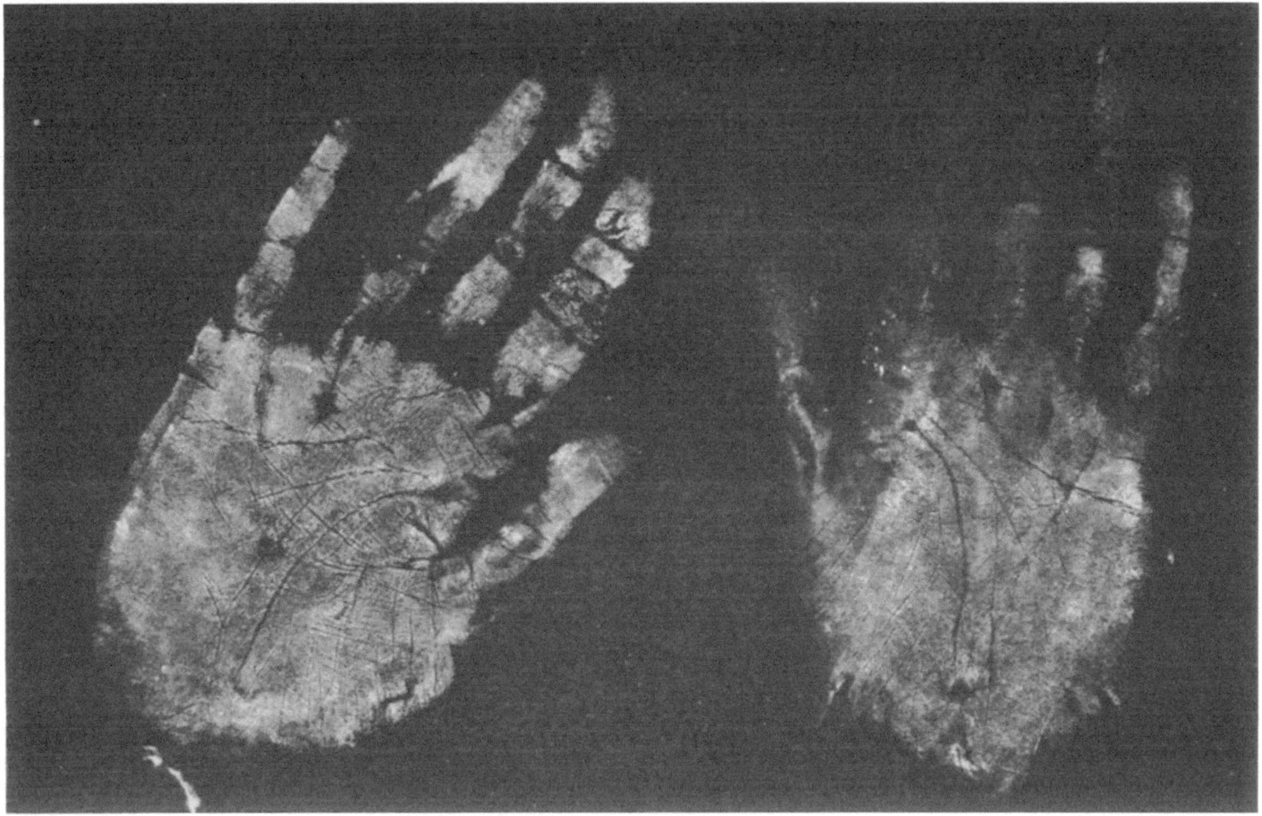

PLATE XLIII

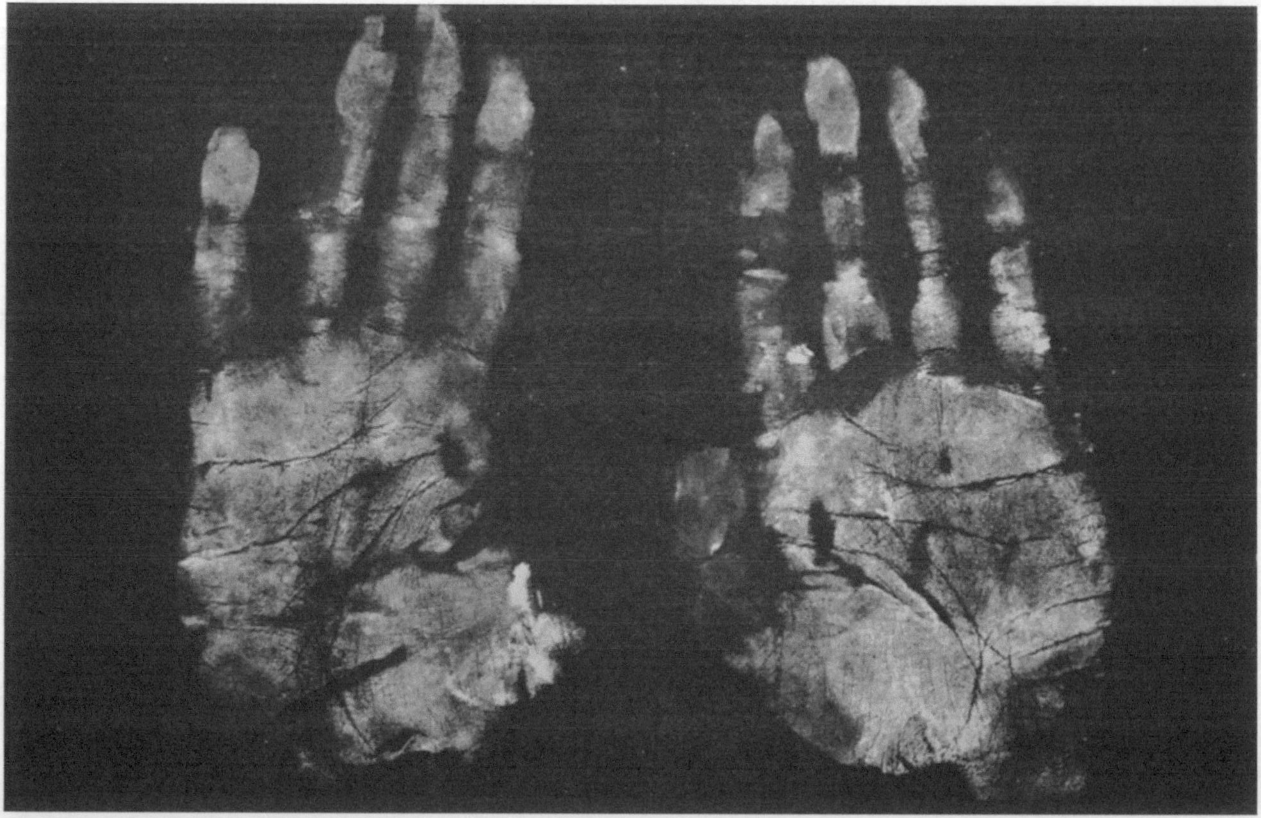

PLATE XLIV

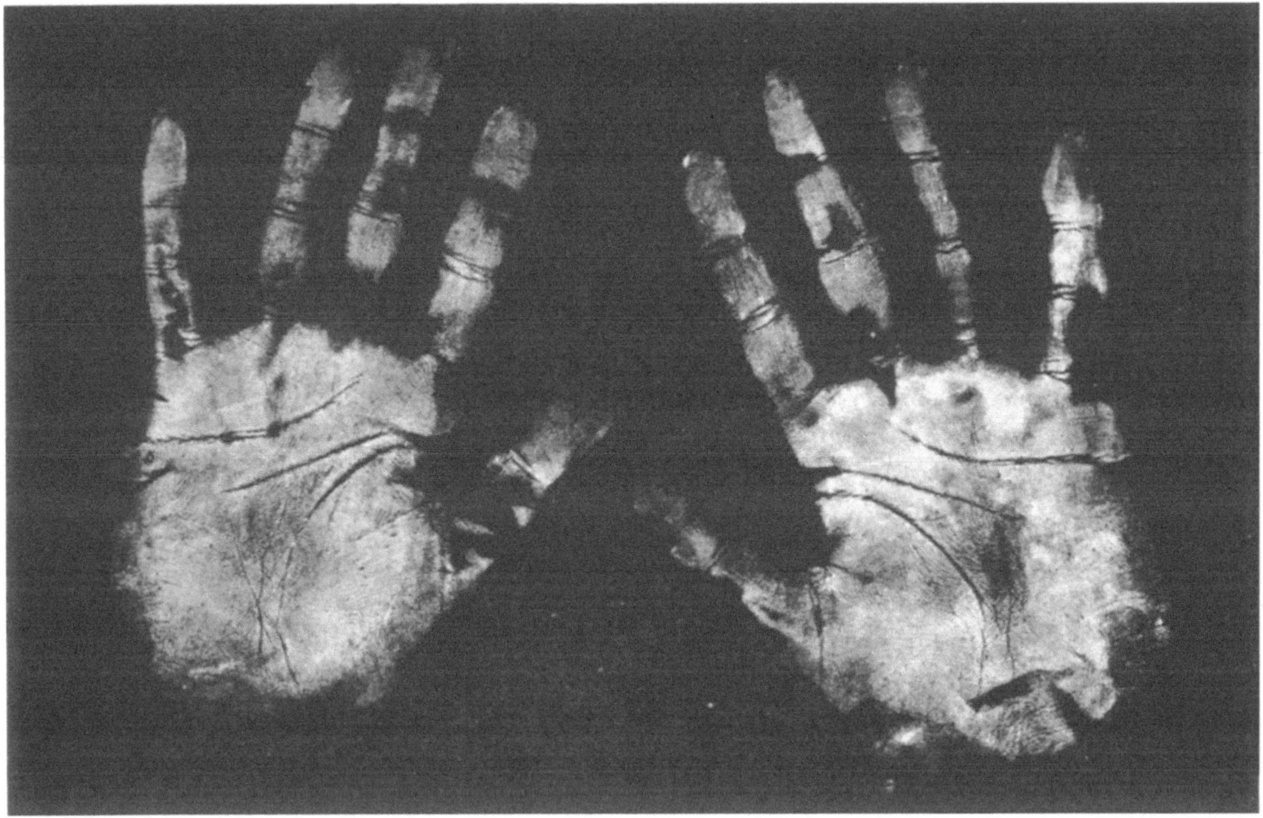

PLATE XLV

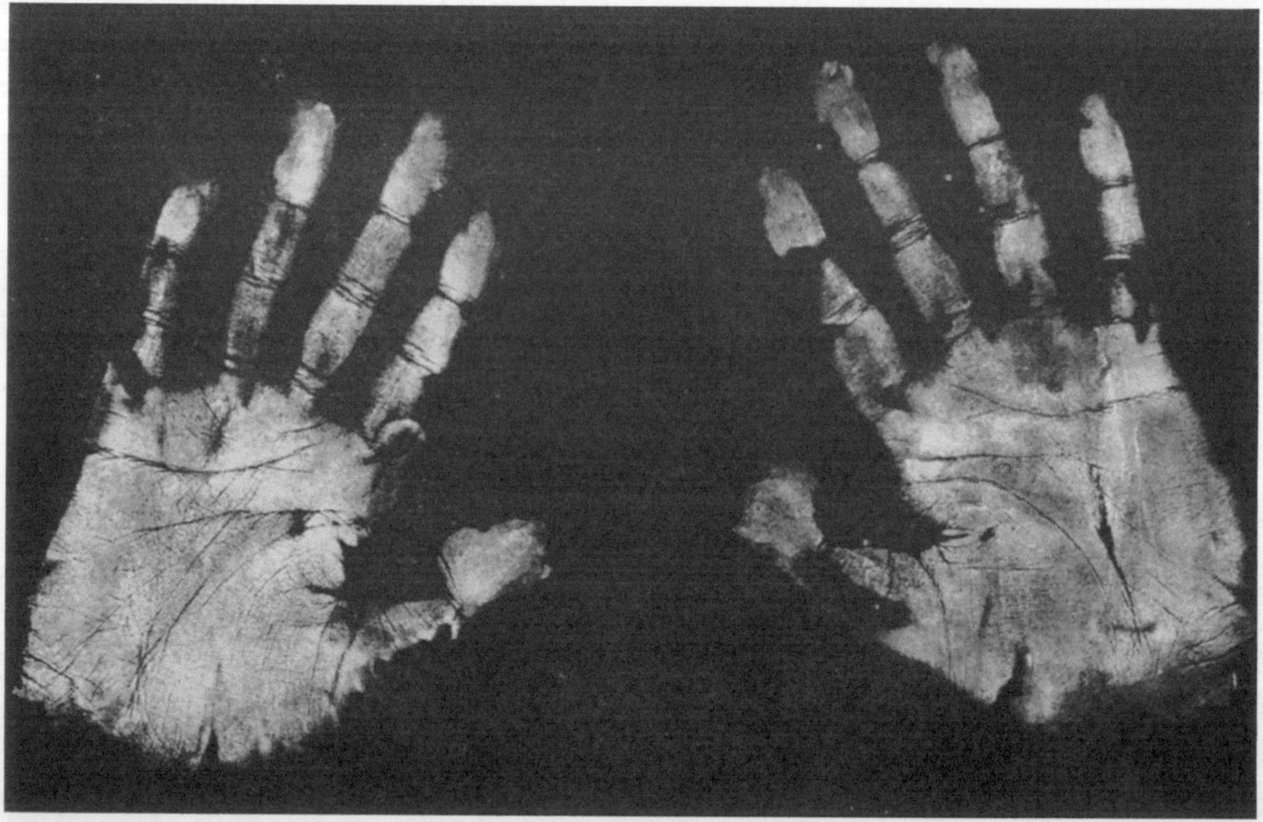

Plate XLVI

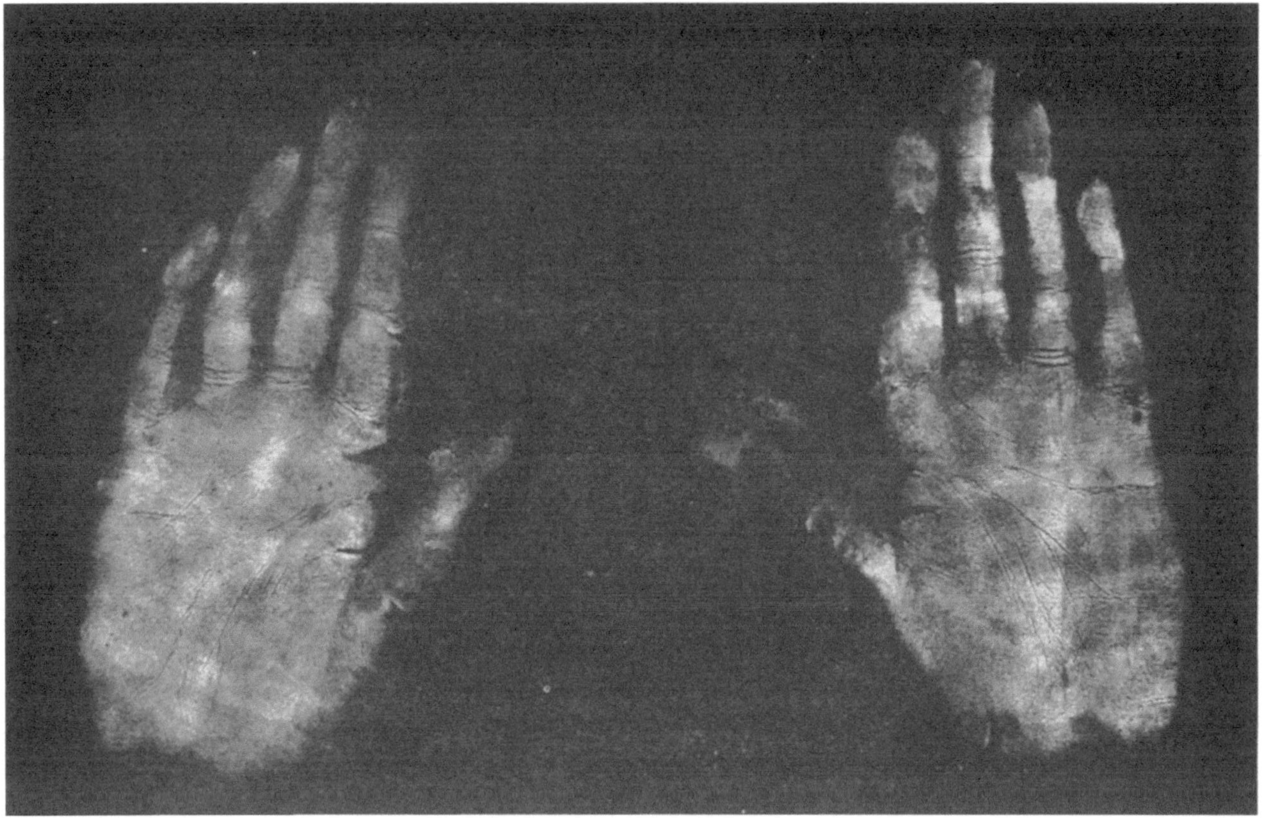

PLATE XLVII

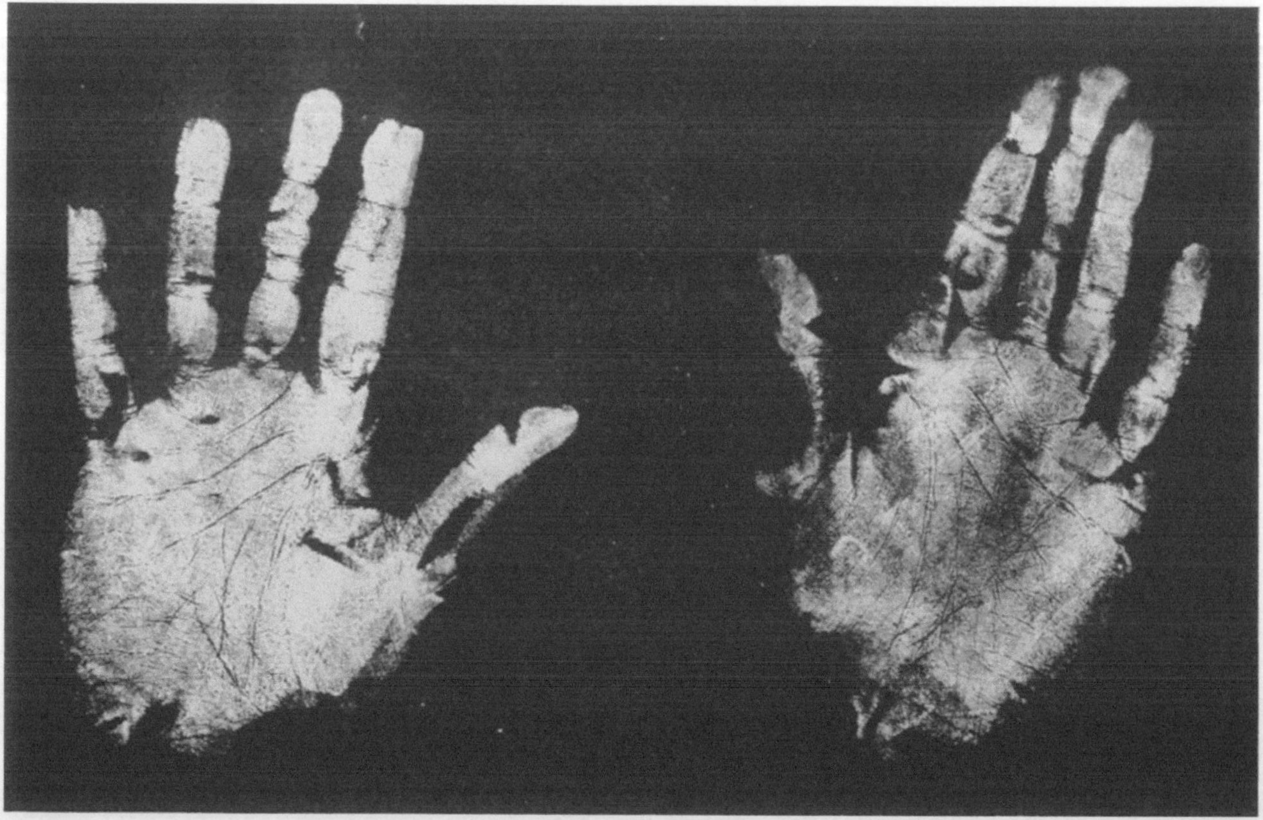

Plate XLVIII

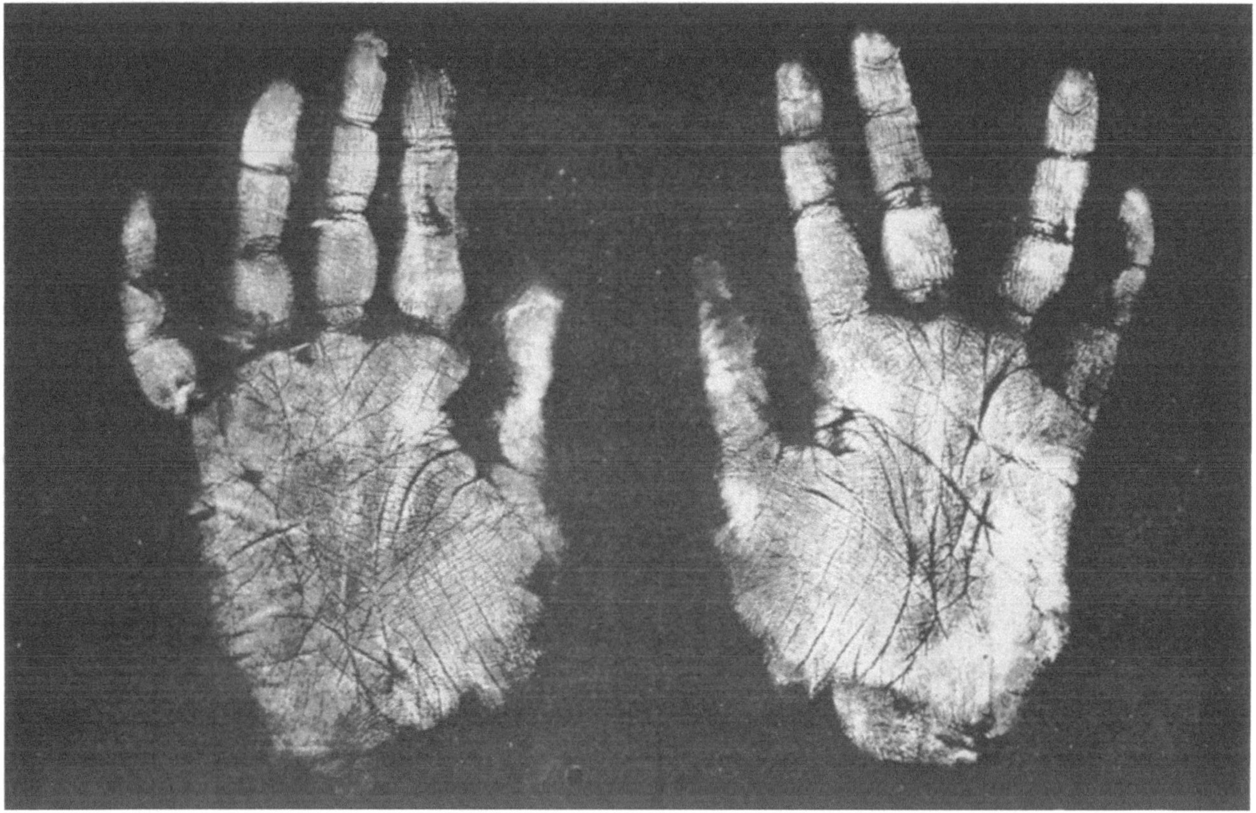

PLATE XLIX

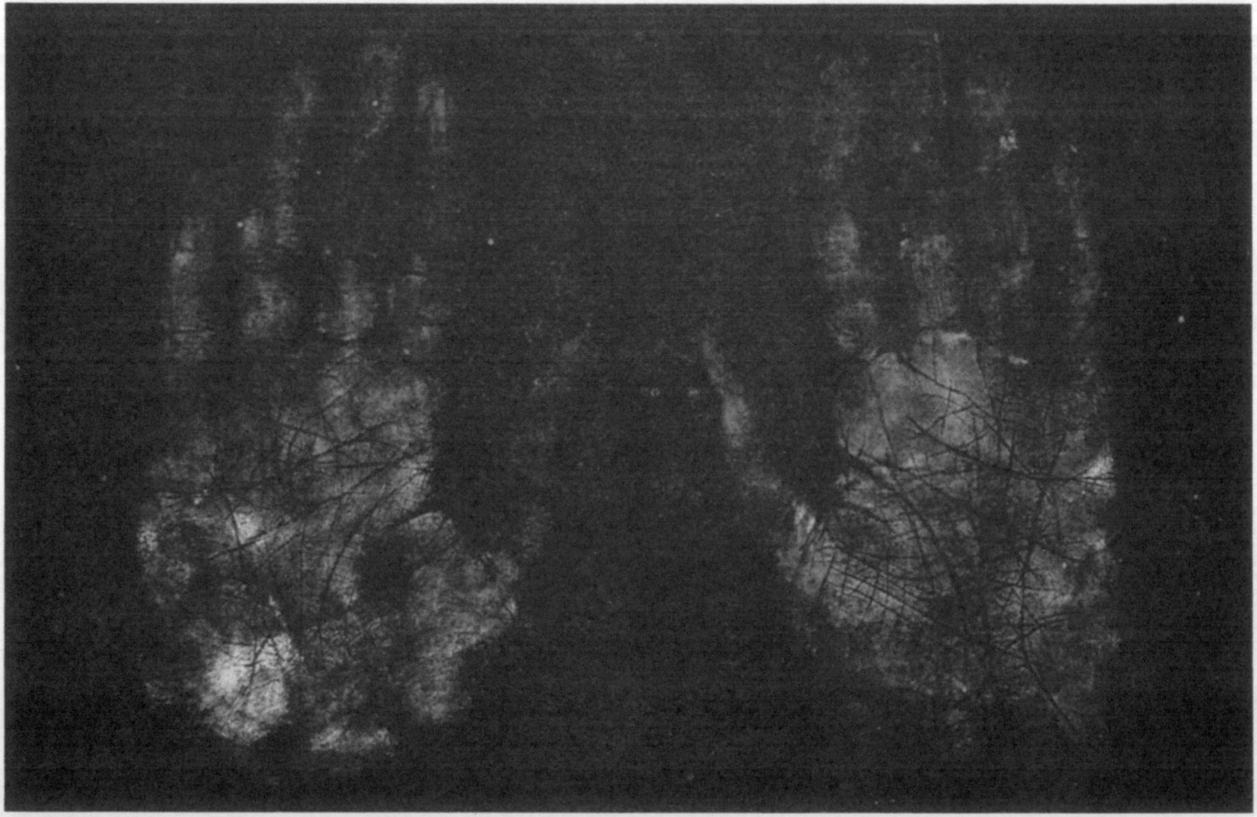

PLATE L

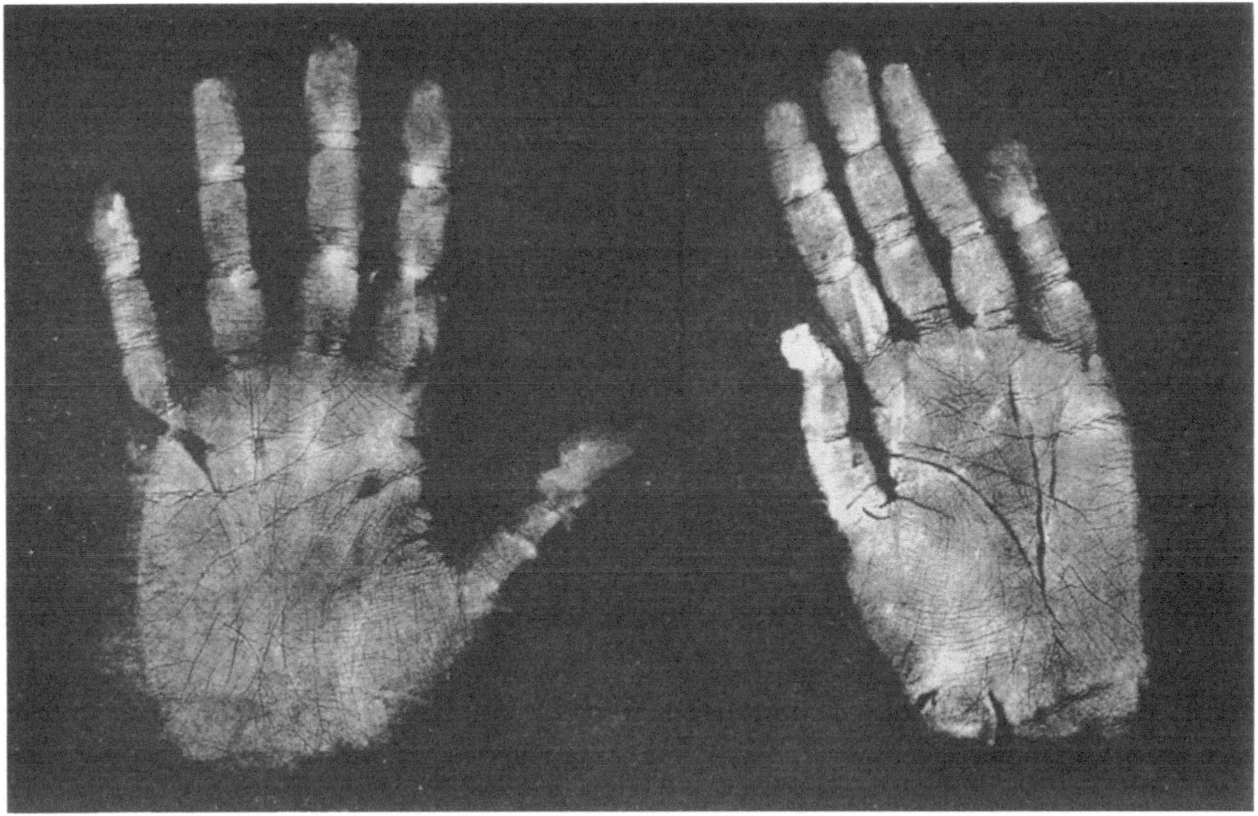

Plate LI

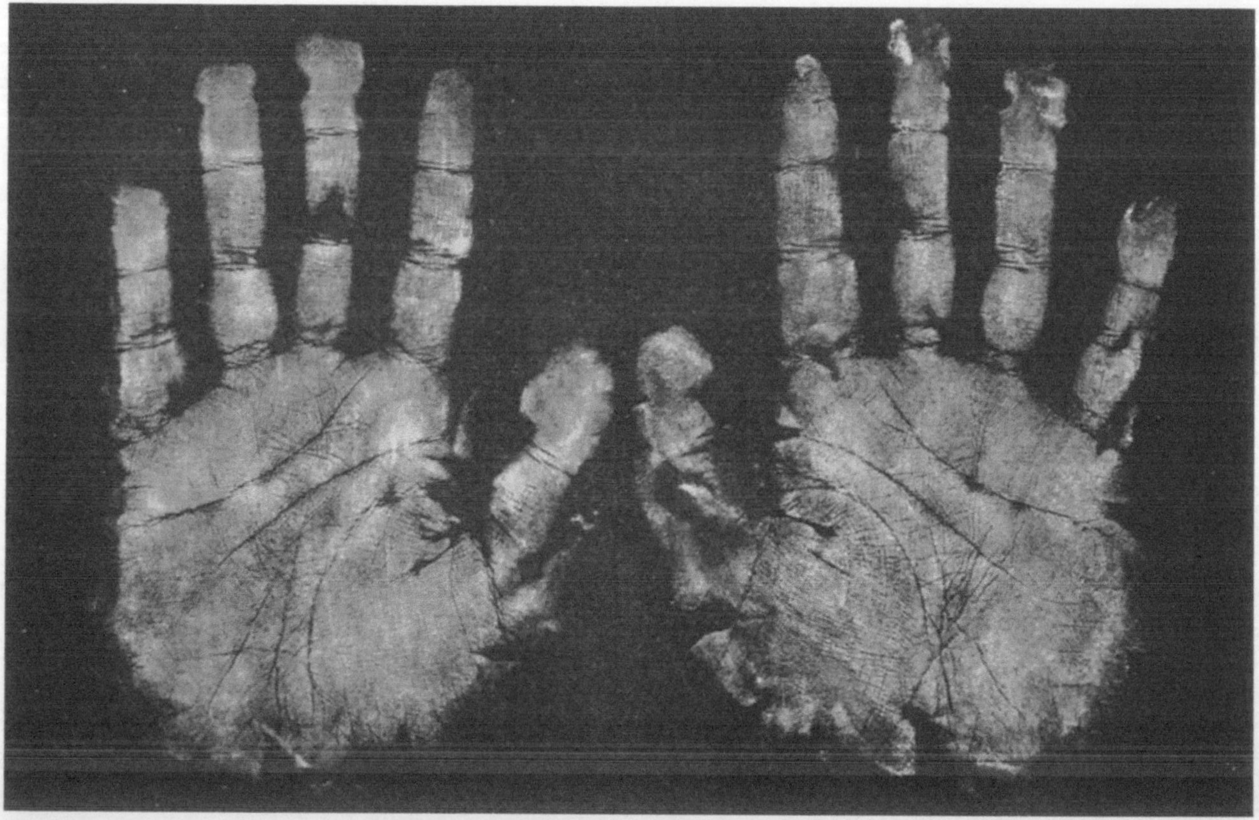

Plate LII

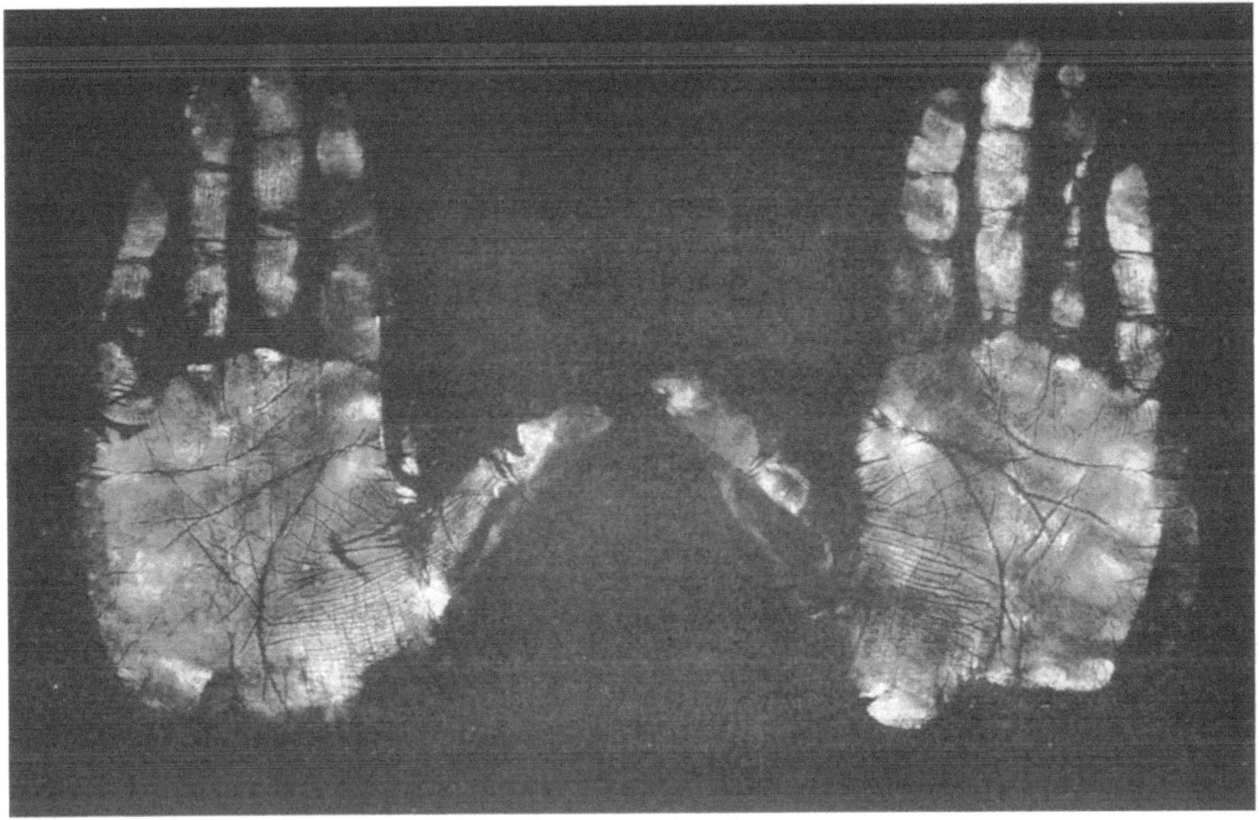

Plate LIII

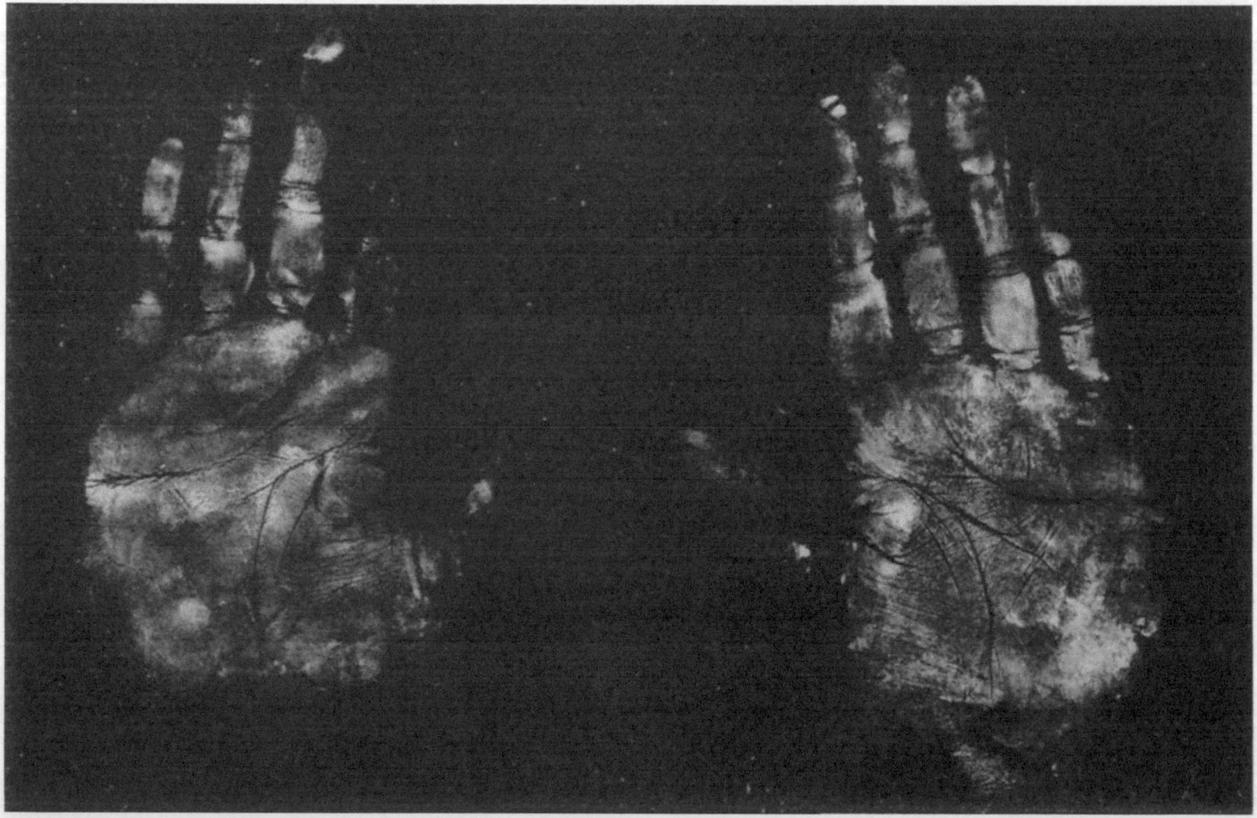

Plate LIV

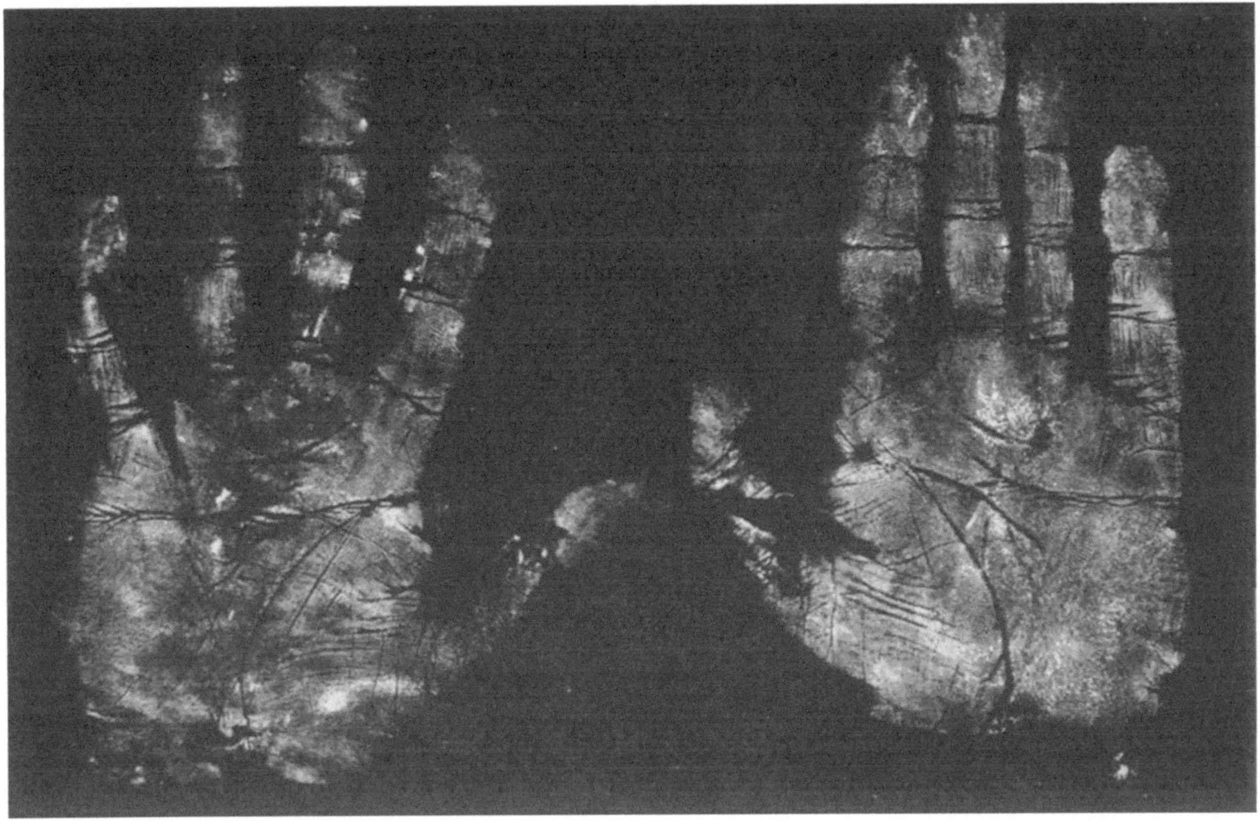

Plate LV

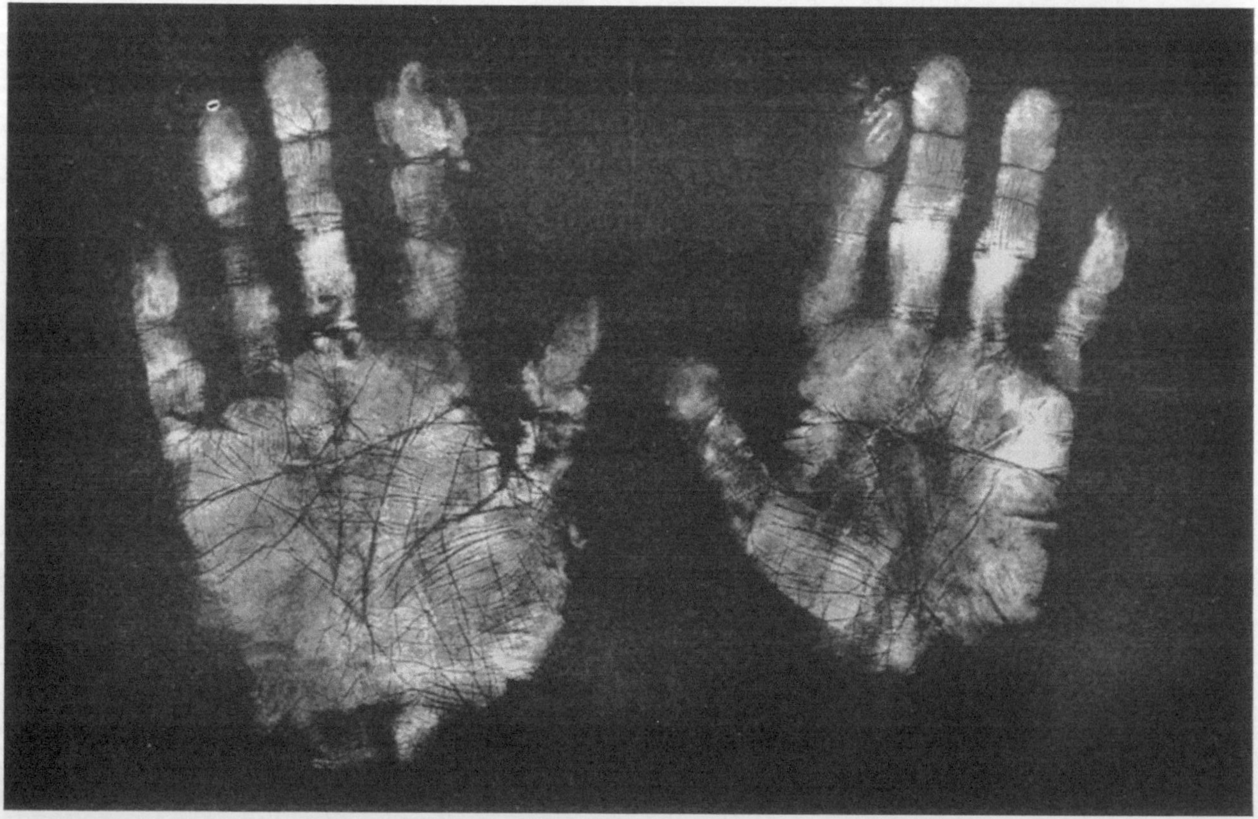

Plate LVI

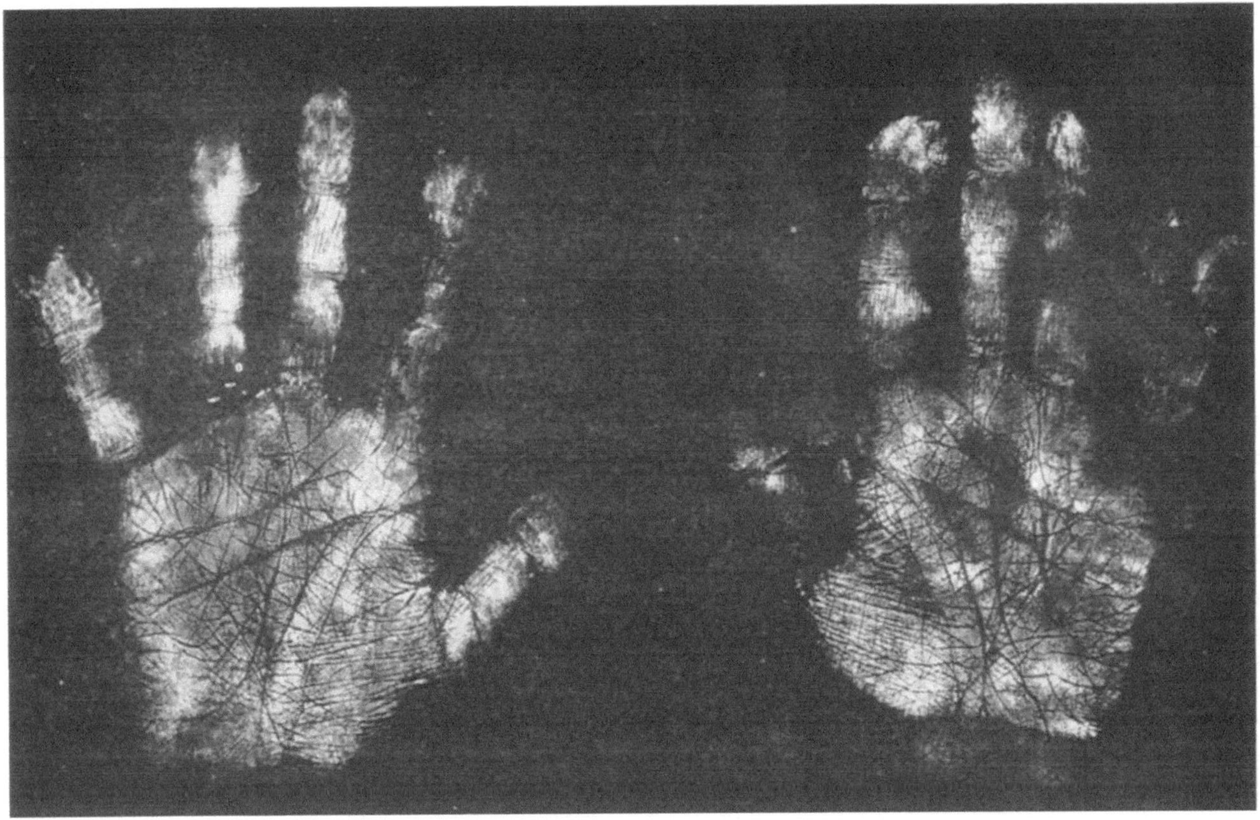

Plate LVII

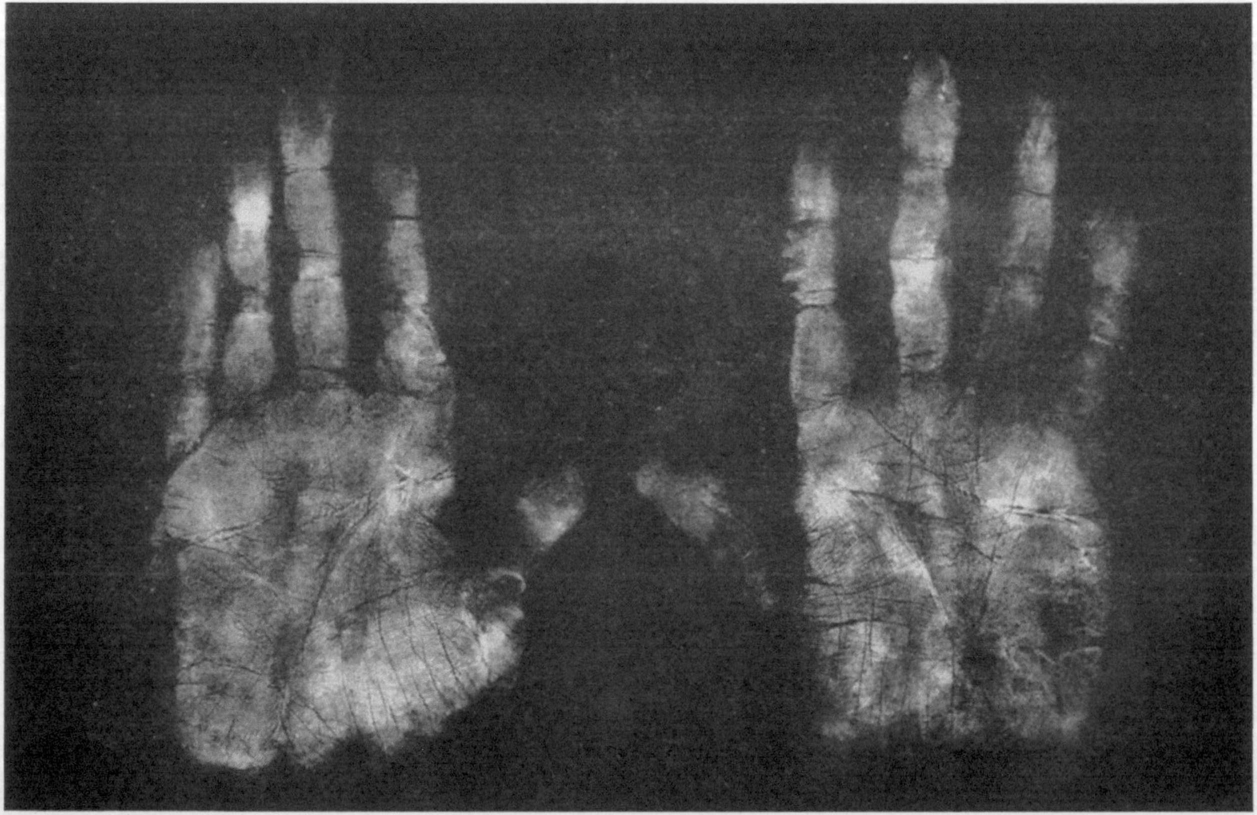

Plate LVIII

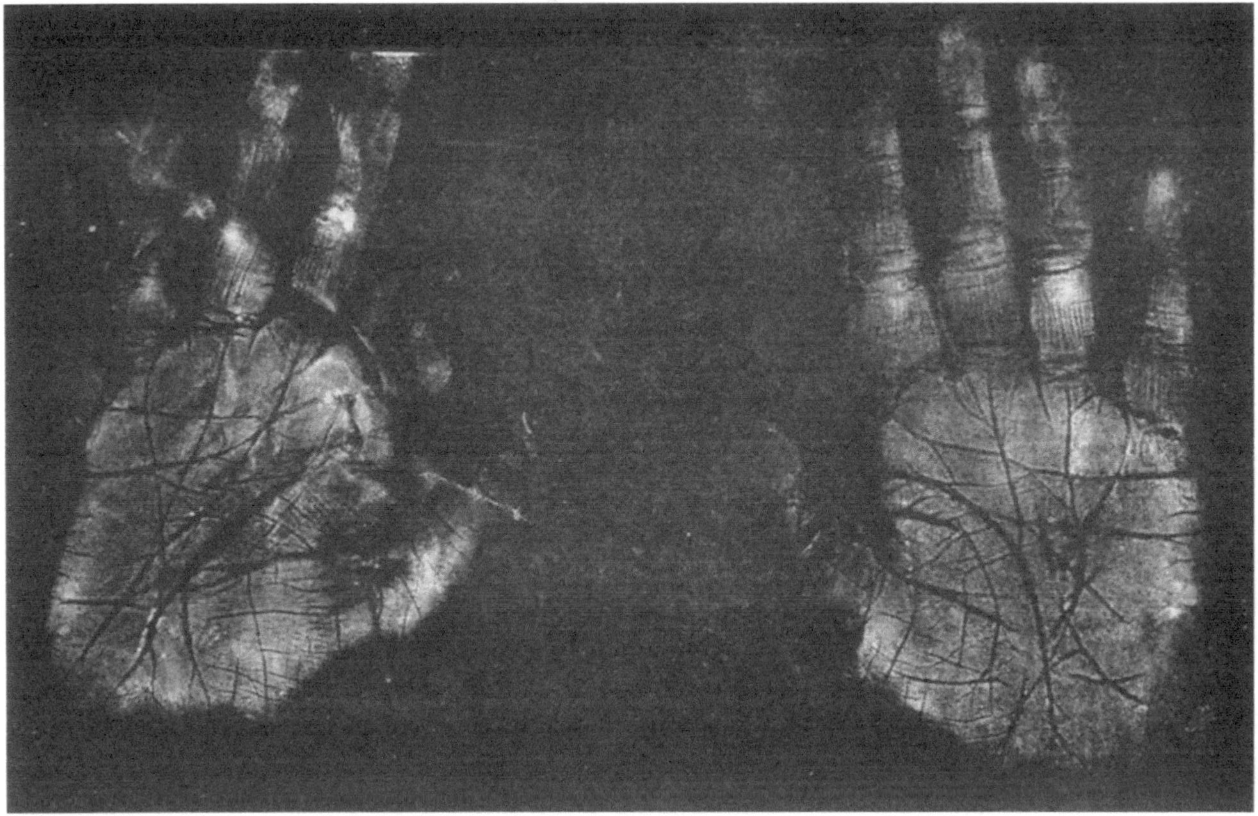

PLATE LIX

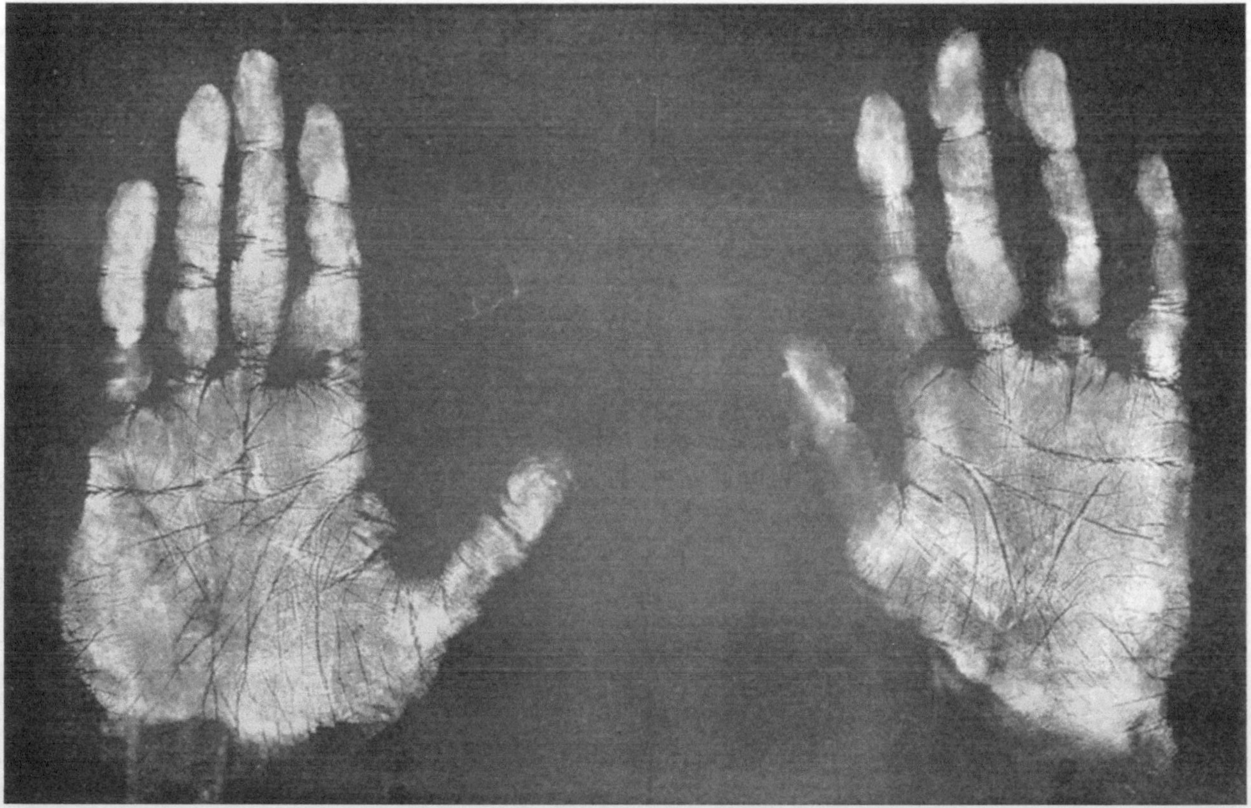

PLATE LX

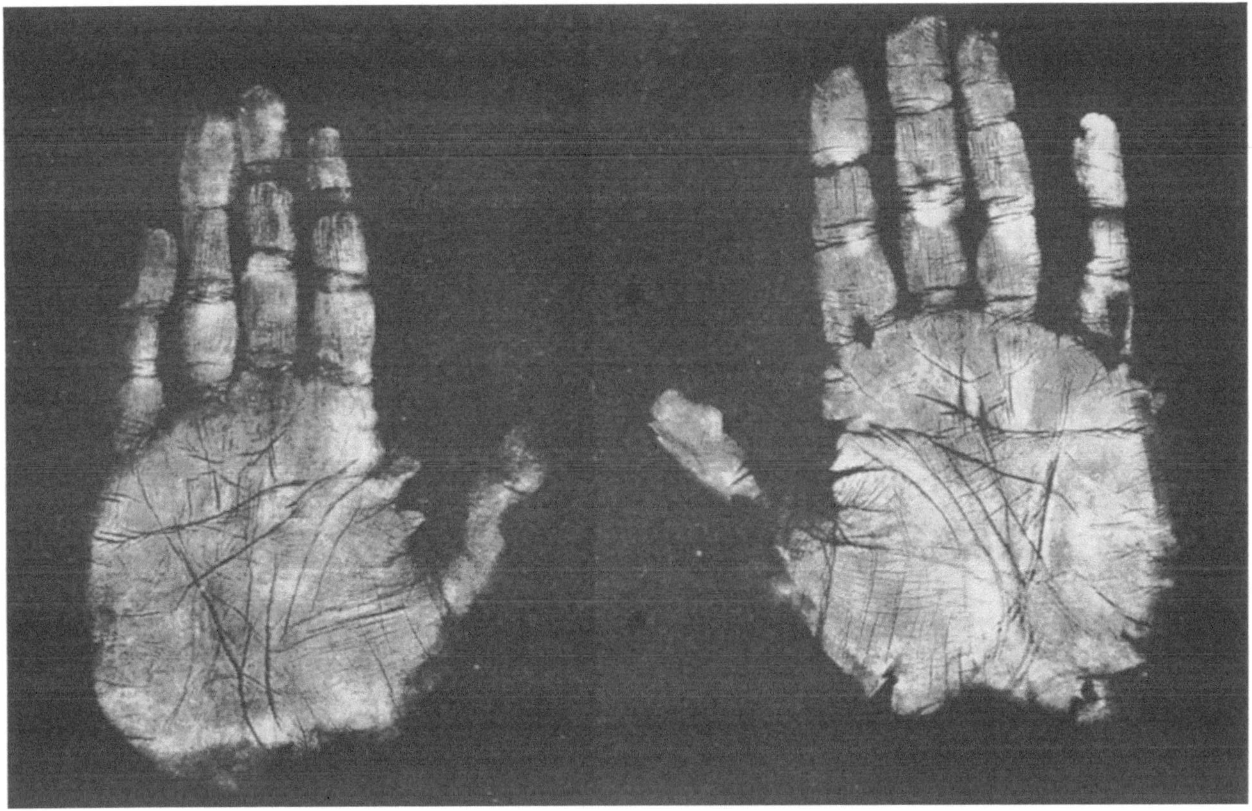

PLATE LXI

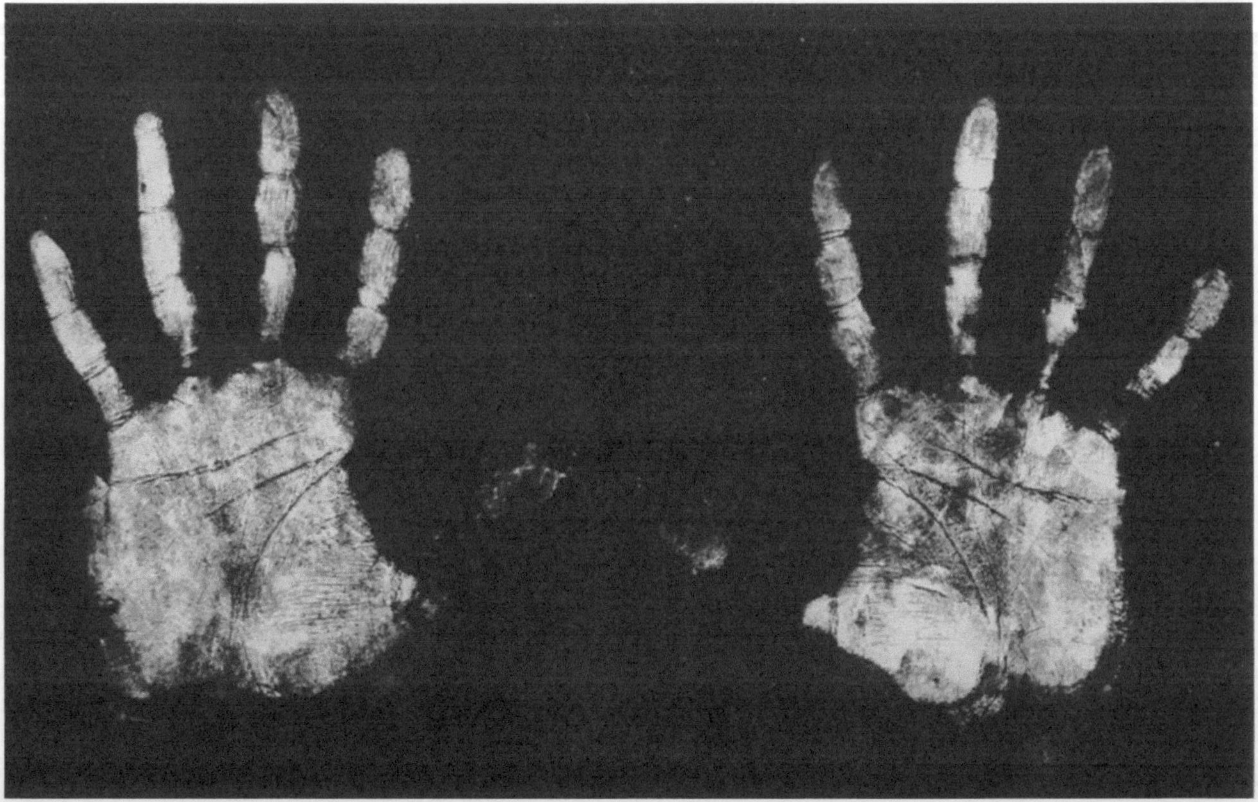

Plate LXII

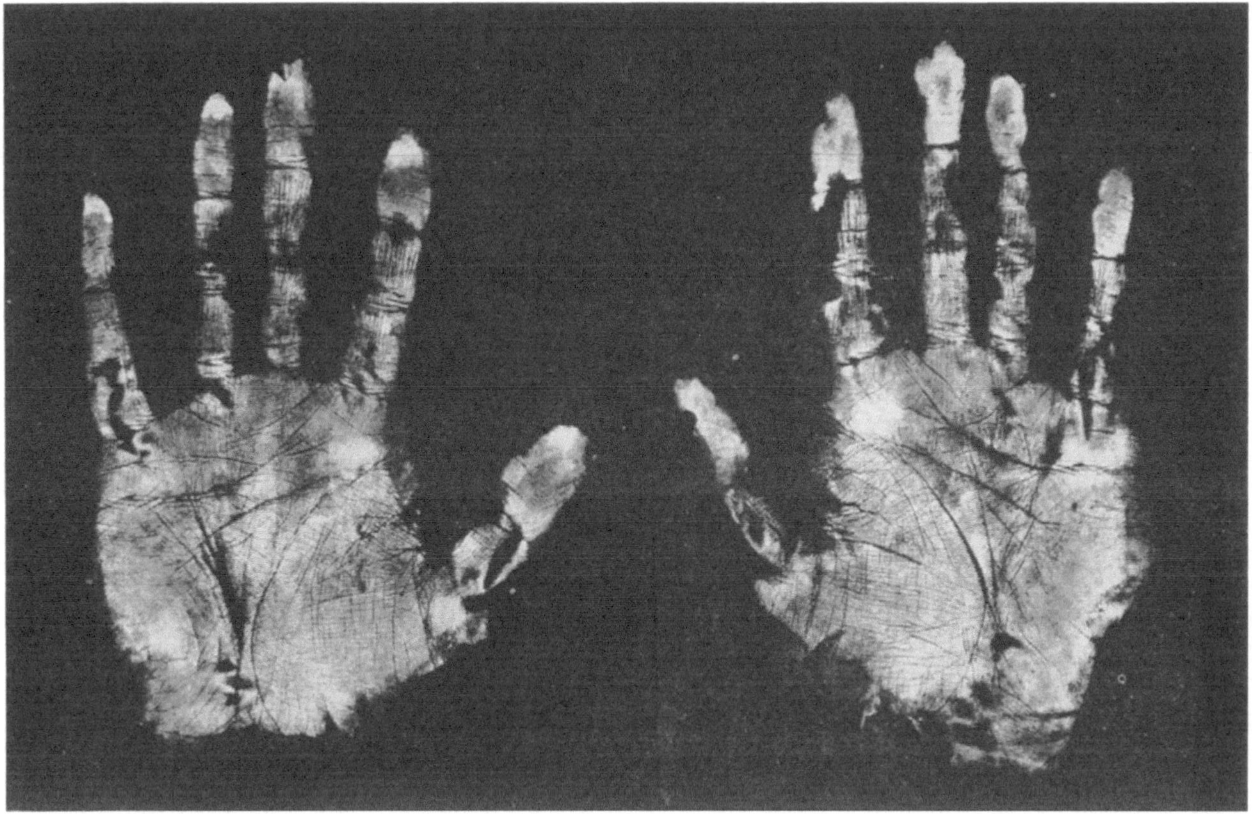

Plate LXIII

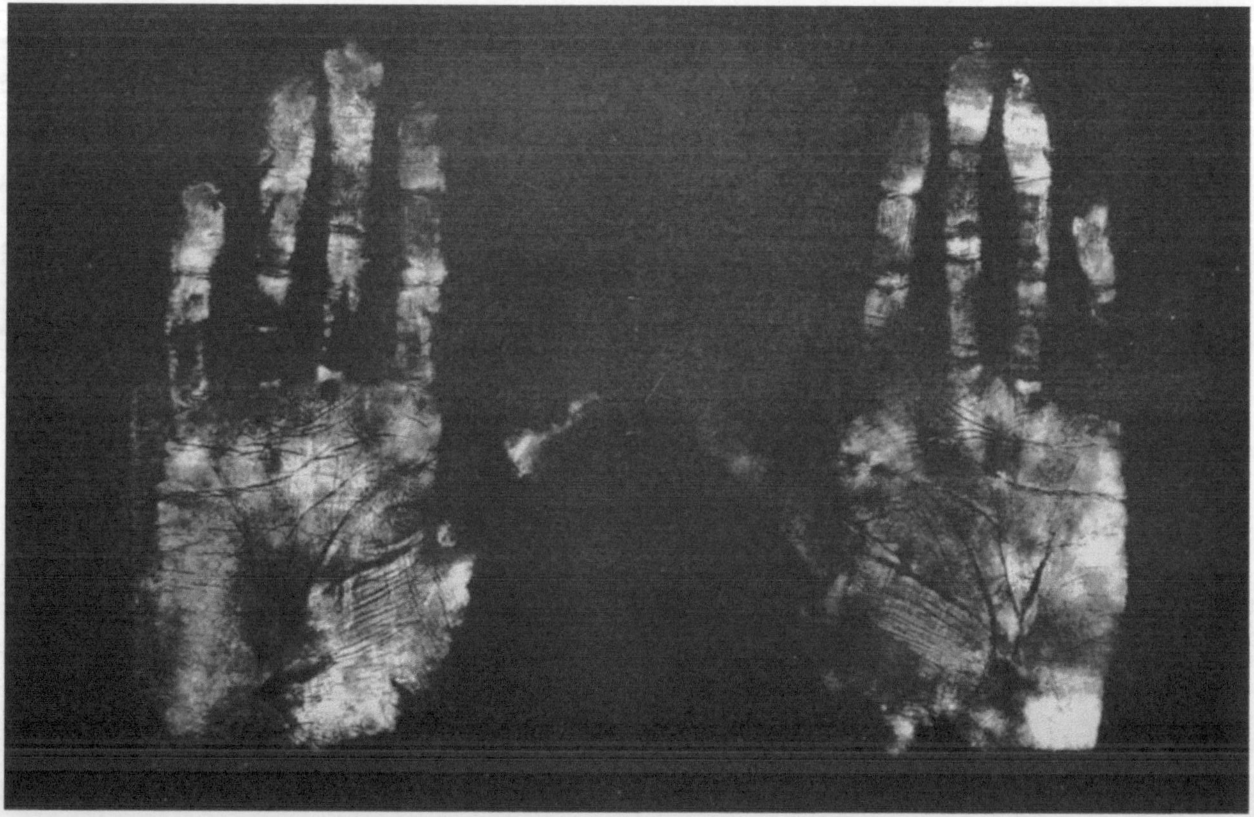

PLATE LXIV

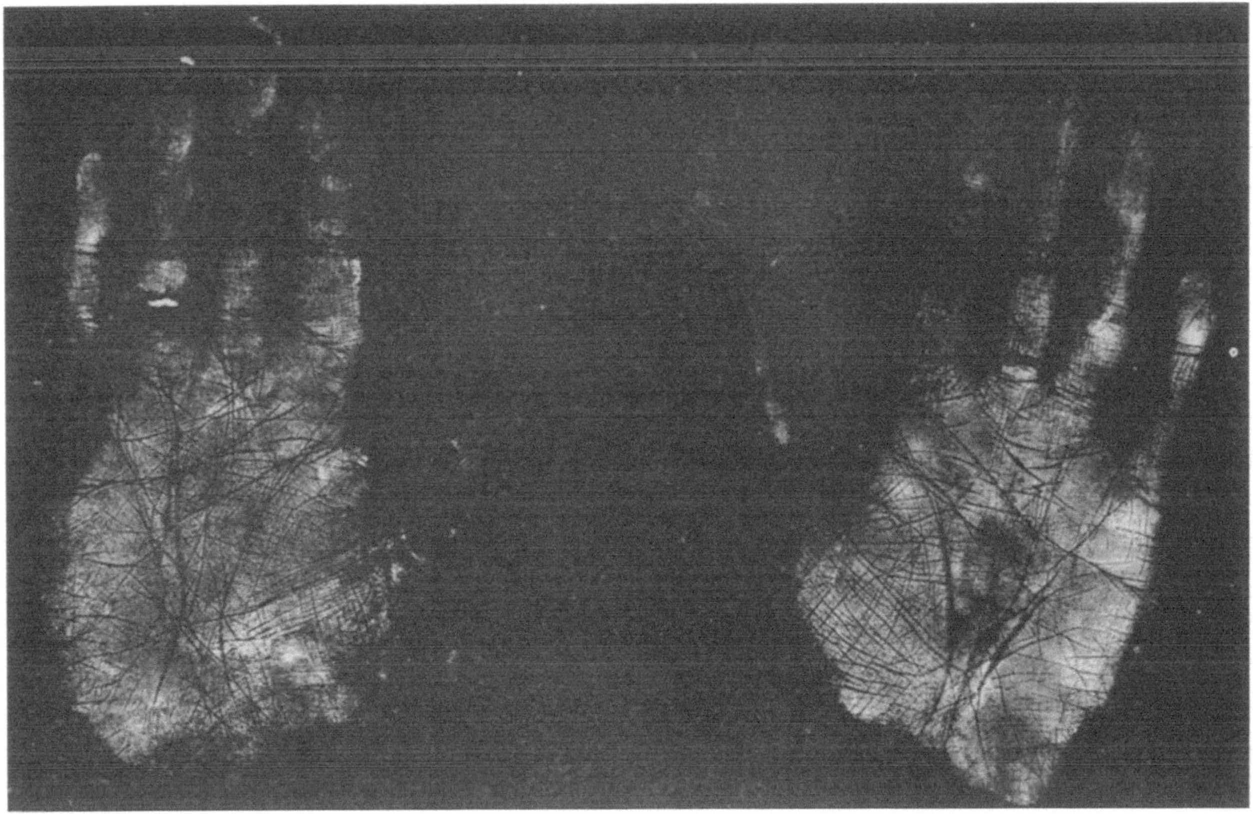

Plate LXV

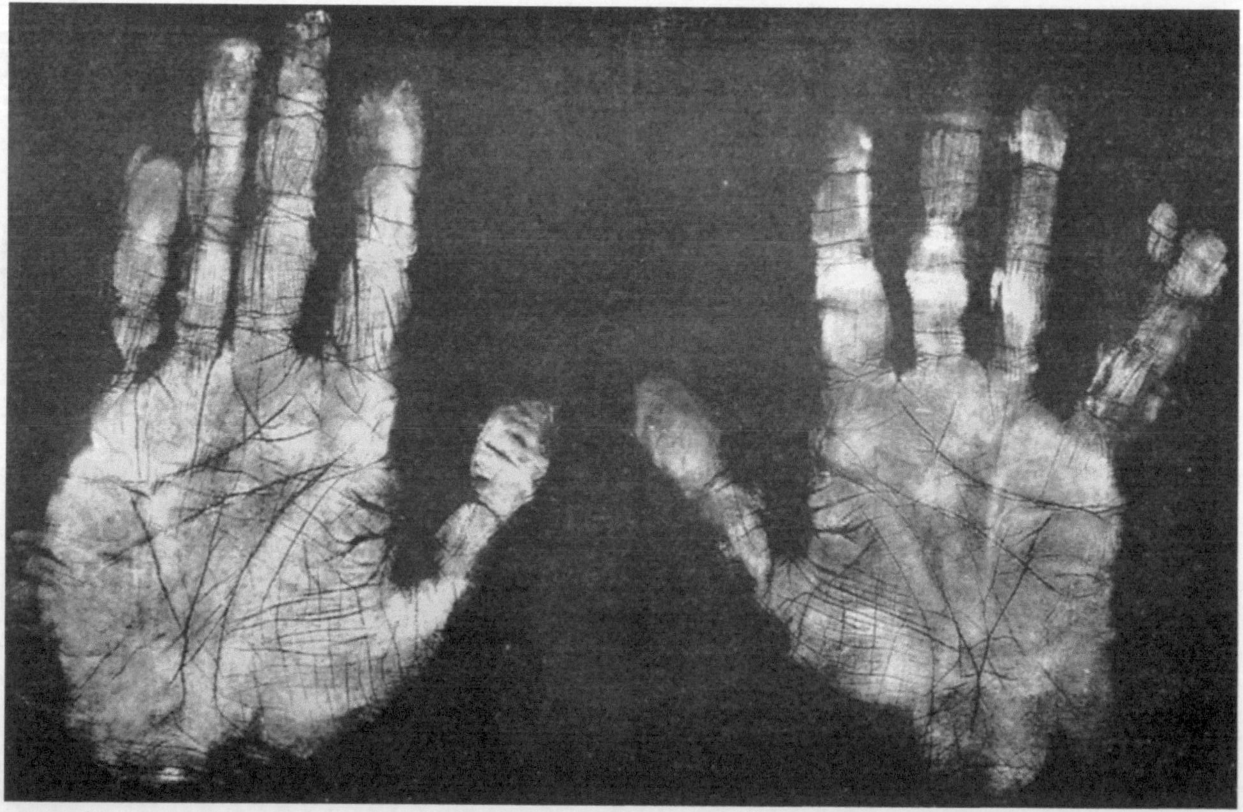

Plate LXVI

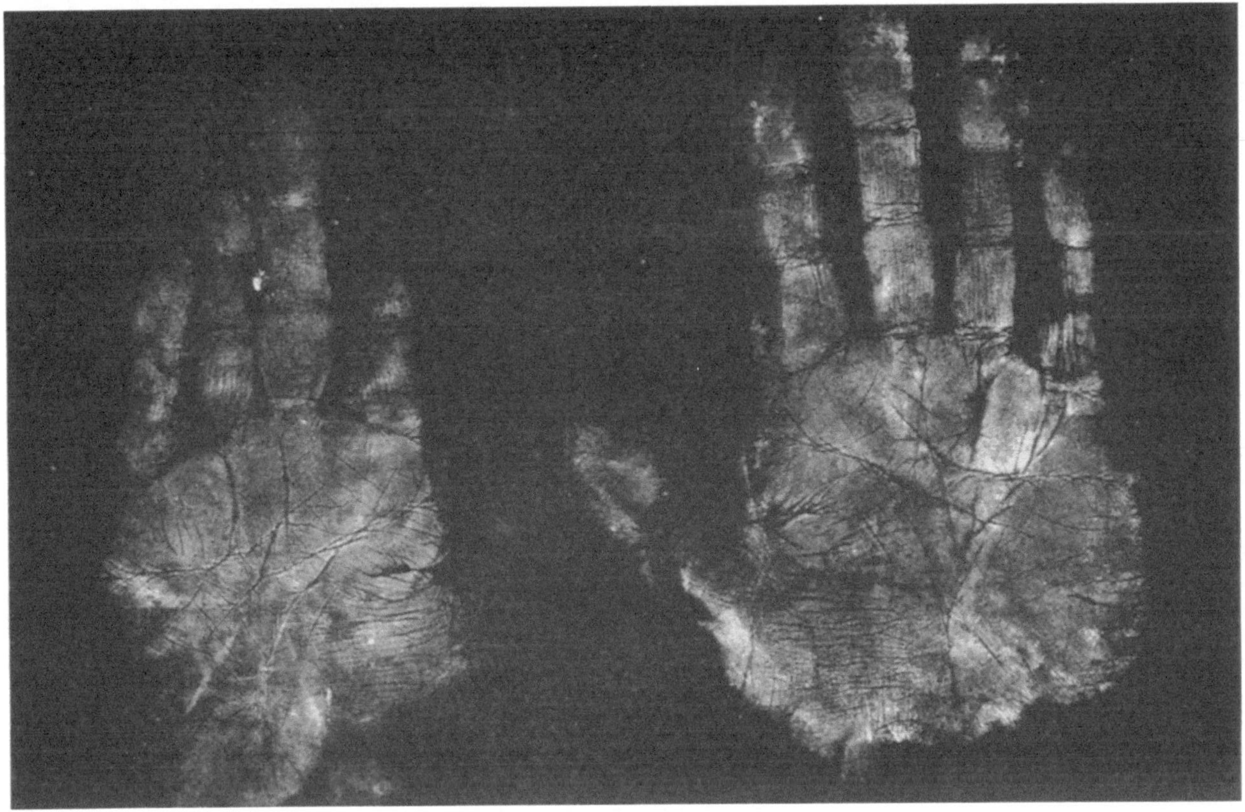

Plate LXVII

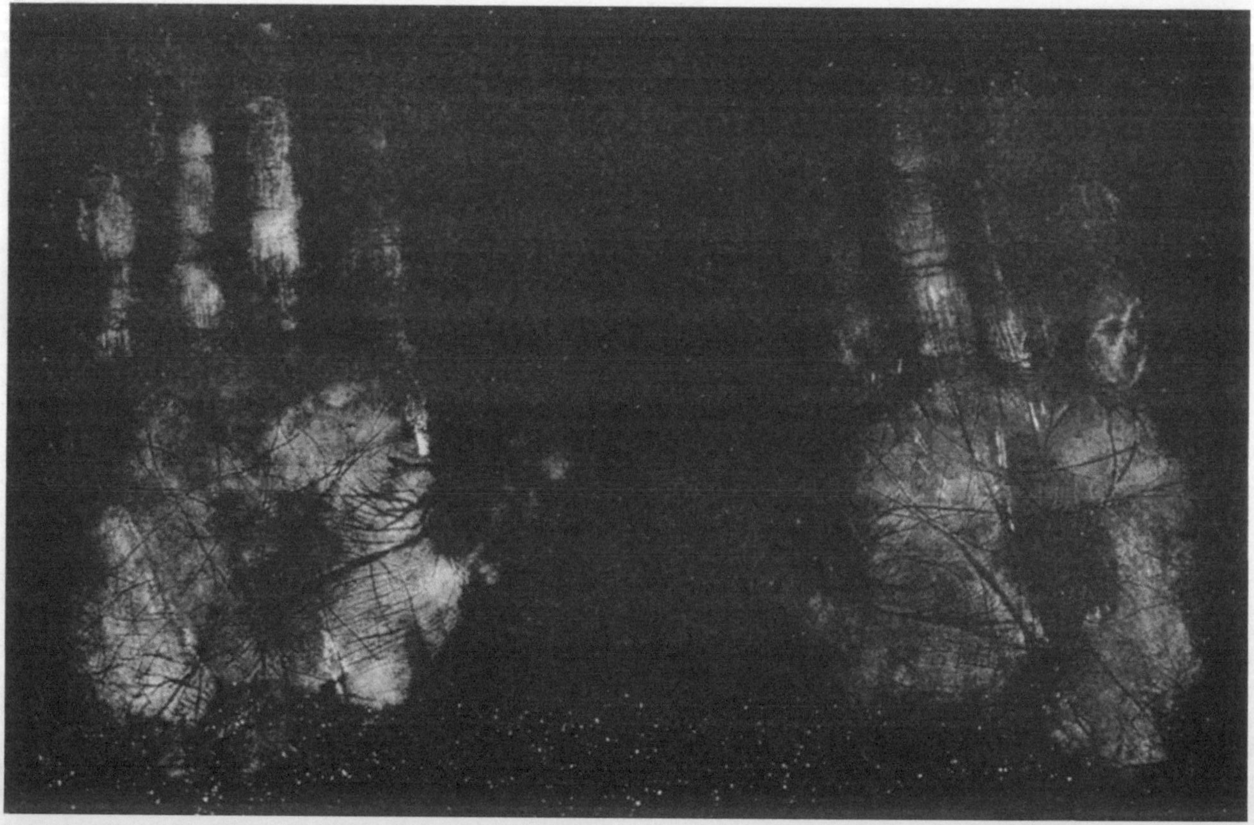

Plate LXVIII

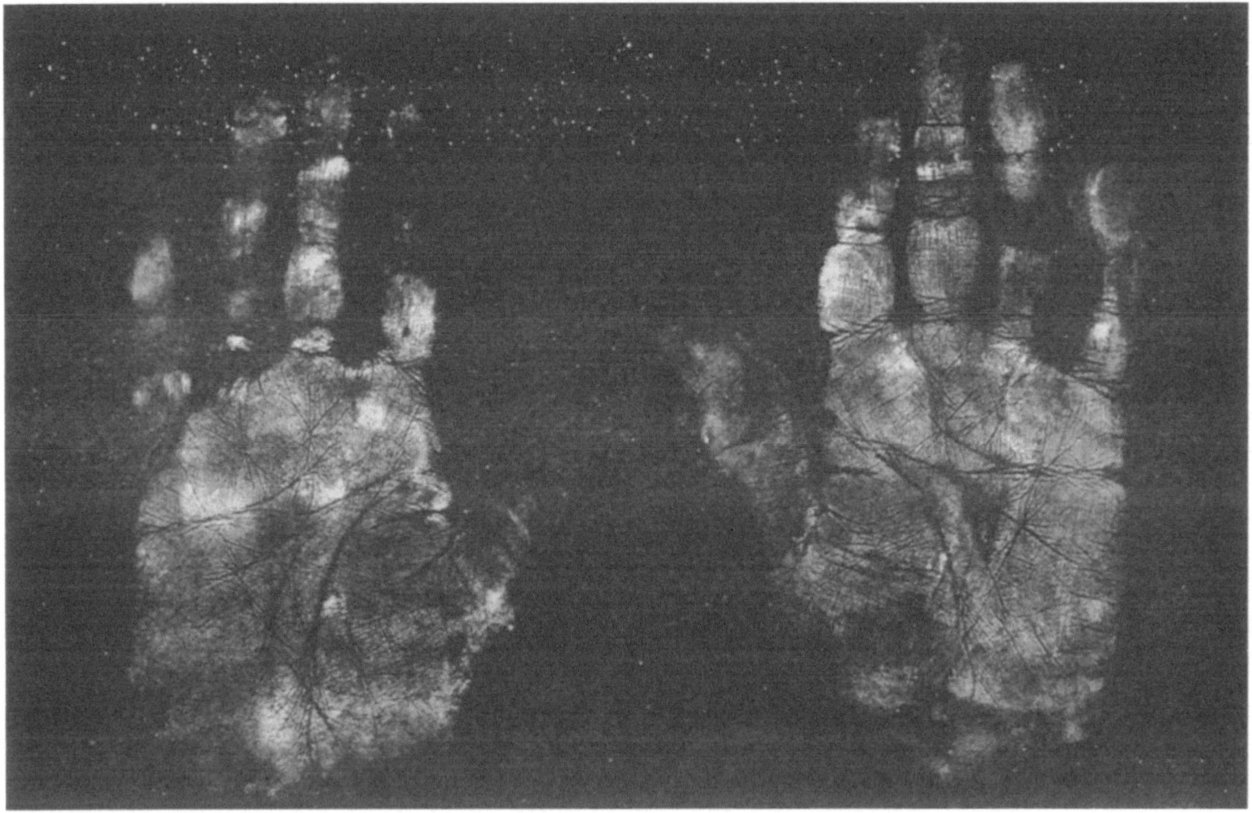

PLATE LXIX

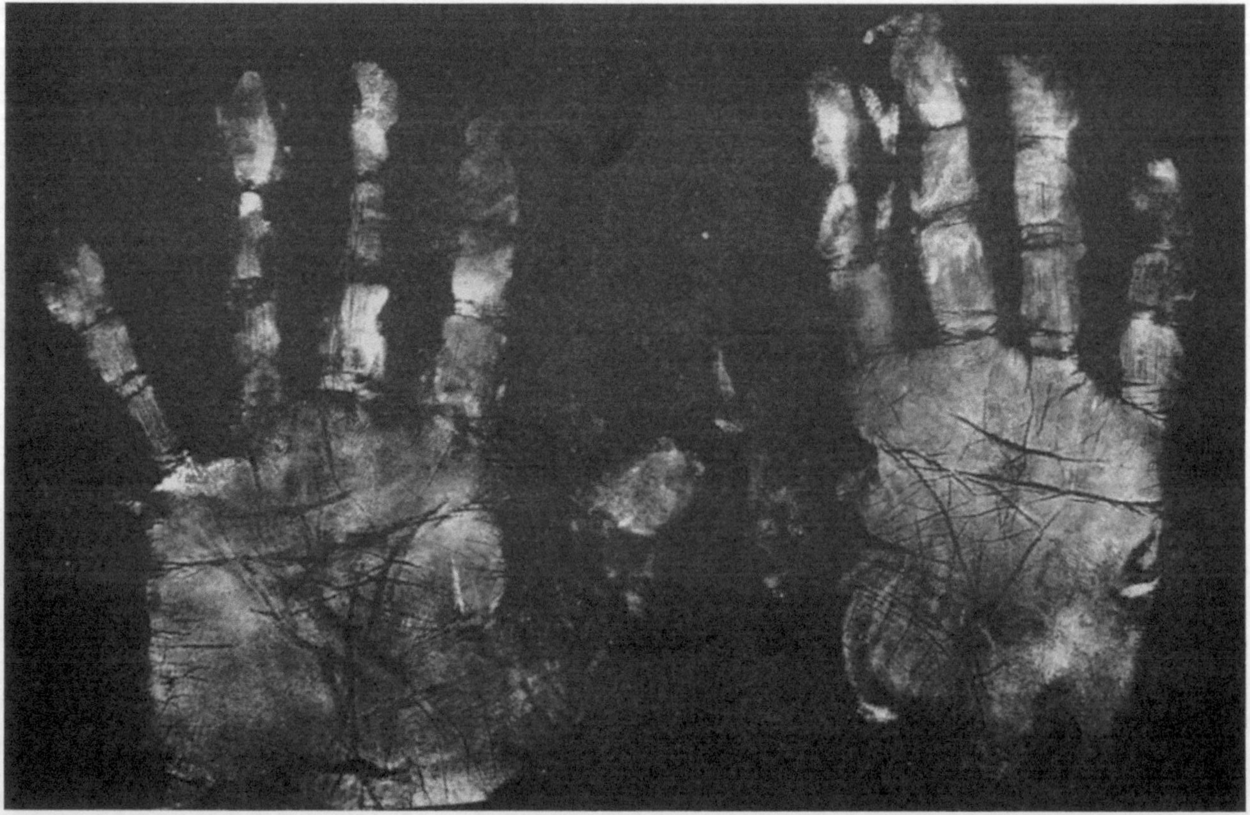